Lecture Notes in Mathematics

Volume 2267

This series reports on new developments in all areas of mathematics and their applications - quickly, informally and at a high level. Mathematical texts analysing new developments in modelling and numerical simulation are welcome. The type of material considered for publication includes:

1. Research monographs
2. Lectures on a new field or presentations of a new angle in a classical field
3. Summer schools and intensive courses on topics of current research.

Texts which are out of print but still in demand may also be considered if they fall within these categories. The timeliness of a manuscript is sometimes more important than its form, which may be preliminary or tentative.

More information about this series at http://www.springer.com/series/304

Tullio Ceccherini-Silberstein • Fabio Scarabotti •
Filippo Tolli

Gelfand Triples and Their Hecke Algebras

Harmonic Analysis for Multiplicity-Free Induced Representations of Finite Groups

Foreword by Eiichi Bannai

Tullio Ceccherini-Silberstein
Dipartimento di Ingegneria
Università degli Studi del Sannio
Benevento, Italy

Fabio Scarabotti
Dipartimento SBAI
Università degli Studi di Roma
"La Sapienza"
Roma, Italy

Filippo Tolli
Dipartimento di Matematica e Fisica
Università degli Studi Roma Tre
Roma, Italy

ISSN 0075-8434 ISSN 1617-9692 (electronic)
Lecture Notes in Mathematics
ISBN 978-3-030-51606-2 ISBN 978-3-030-51607-9 (eBook)
https://doi.org/10.1007/978-3-030-51607-9

Mathematics Subject Classification: 20C15, 43A65, 20C08, 20G05, 43A90, 43A35

This Springer imprint is published by the registered company Springer Nature Switzerland AG.
The registered company address is: Gewerbestrasse 11, 6330 Cham, Switzerland

To Francesca
To the memory of my father
To Valentina

Foreword

A (finite) Gelfand pair is a pair (G, H) consisting of a (finite) group G and a subgroup H such that G, with its action on the coset space G/H, yields a multiplicity-free transitive permutation group, equivalently, the induced representation $\mathrm{Ind}_H^G(1_H)$ of the identity representation 1_H of H to G is multiplicity-free, i.e., decomposes into pairwise-inequivalent irreducible representations of G.

Gelfand pairs constitute an important mathematical concept that plays a central role in several areas of mathematics. The purely combinatorial counterpart of Gelfand pairs are the (commutative) association schemes. Gelfand pairs, indeed, constitute the most important class of (commutative) association schemes. Both Gelfand pairs and association schemes have already been extensively studied, because of their importance in their own right. But there are many other objects and theories that have been studied and developed under this framework, for example, distance-transitive and/or distance-regular graphs, coding theory and design theory (Delsarte theory), probability theory and statistics (Diaconis theory), witnessing, once more, the strong connections of Gelfand pairs and association schemes with many other areas of mathematics.

There have been many attempts to generalize the concept of a Gelfand pair. Some authors have already successfully dealt with the case of a triple (G, H, η), where G and H are a group and a subgroup as before, and η is a linear representation of H such that $\mathrm{Ind}_H^G(\eta)$ is multiplicity-free.

The next more general step would naturally be to study the case of a triple (G, H, η), where now η is an arbitrary (not necessarily linear) irreducible representation of H such that $\mathrm{Ind}_H^G(\eta)$ is multiplicity-free. The authors of this memoir call (G, H, η) a Gelfand triple. It seems that this situation has been regarded by many experts to be a bit too general, and this may explain why not much general theory has been developed so far.

The present memoir tackles this problem very seriously and bravely from the very front side. Frankly speaking, in spite of some success achieved in this work, many questions are still left unanswered and waiting to be studied. However, I believe that this text provides us with very useful foundations and information to start a serious research in this direction. This memoir is "the" pioneering work on general Gelfand triples.

The three authors are very strong researchers working in representation theory and discrete harmonic analysis, as well as in many related fields of mathematics. They published already several excellent books on these topics. This memoir has a mixed nature of both a research paper and a book. Indeed, on the one hand, all the detailed proofs are carefully given as in a research paper, and, on the other hand, the authors masterfully describe the mathematical philosophy of this research direction, as witnessed in any excellent book.

I believe that this volume in the Springer LNM series will provide another good addition to this general research direction, namely "harmonic analysis on finite groups." It will be read and very welcomed, not only by experts but also by a broad range of mathematicians.

Tokyo, Japan Eiichi Bannai
May 2020

Preface

Finite Gelfand pairs play an important role in mathematics and have been studied from several points of view: in algebra (we refer, for instance, to the work of Bump and Ginzburg [7, 8] and Saxl [57]; see also [15]), in representation theory (as witnessed by the new approach to the representation theory of the symmetric groups by Okounkov and Vershik [54], see also [13]), in analysis (with relevant contributions to the theory of special functions by Dunkl [30] and Stanton [67]), in number theory (we refer to the book by Terras [69] for a comprehensive introduction; see also [11, 17]), in combinatorics (in the language of association schemes as developed by Bannai and Ito [1]), and in probability theory (with the remarkable applications to the study of diffusion processes by Diaconis [24]; see also [10, 11]). Indeed, Gelfand pairs arise in the study of algebraic, geometrical, or combinatorial structures with a large group of symmetries such that the corresponding permutation representations decompose without multiplicities: it is then possible to develop a useful theory of spherical functions with an associated spherical Fourier transform.

In our preceding work, we have shown that the theory of spherical functions may be studied in a more general setting, namely for permutation representations that decompose with multiplicities [14, 58], for subgroup-conjugacy-invariant functions [16, 59], and for general induced representations [61]. Indeed, a finite Gelfand pair may be considered as the simplest example of a multiplicity-free induced representation (the induction of the trivial representation of the subgroup), and this is the motivation for the present monograph.

The most famous of these multiplicity-free representations is the Gelfand–Graev representation of a reductive group over a finite field [34] (see also Bump [7]). In this direction, we have started our investigations in Part IV of our monograph [17], where we have developed a theory of spherical functions and spherical representations for multiplicity-free induced representations of the form $\mathrm{Ind}_K^G \chi$, where χ is a one-dimensional representation of subgroup K. This case was previously investigated by Stembridge [68], Macdonald [46, Exercise 10, Chapter VII], and Mizukawa [49, 50]. We have applied this theory to the Gelfand–Graev representation of $\mathrm{GL}(2, \mathbb{F}_q)$, following the beautiful expository paper of Piatetski-Shapiro [53], where the author did not use the terminology/theory of spherical functions but, actually, computed them. In such a way, we have shed light on the results and the calculations in [53] by framing them in a more comprehensive theory.

In the present monograph, we face the more general case: we study multiplicity-free induced representations of the form $\mathrm{Ind}_K^G\theta$, where θ is an irreducible K-representation, not necessarily one-dimensional. In this case, borrowing a terminology used by Bump in [7, Section 47], we call (G, K, θ) a *multiplicity-free triple*. Since this constitutes a generalization of Gelfand pairs, we shall also refer to (G, K, θ) as to a *Gelfand triple*, although we are aware that such a terminology is already widely used in another setting, namely in functional analysis and quantum mechanics, as a synonym of a *rigged Hilbert space* [23, 35].

Our first target (cf. Sect. 2.1) is a deep analysis of Mackey's formula for invariants. We show that the commutant $\mathrm{End}_G(\mathrm{Ind}_K^G\theta)$ of an arbitrary induced representation $\mathrm{Ind}_K^G\theta$, with θ an irreducible K-representation, is isomorphic to both a suitable convolution algebra of operator-valued functions defined on G and to a subalgebra of the group algebra of G. We call it the *Hecke algebra* of the triple (G, K, θ) (cf. Bump [7, Section 47], Curtis and Reiner [22, Section 11D], and Stembridge [68]; see also [17, Chapter 13] and [60]). Note that this study does not assume multiplicity-freeness. In fact, we shall see (cf. Theorem 3.1) that the triple (G, K, θ) is multiplicity-free exactly when the associated Hecke algebra is commutative.

We then focus on our main subject of study, namely multiplicity-free induced representations (cf. Chap. 3); we extend to higher dimensions a criterion of Bump and Ginzburg from [8]: this constitutes an analogue of the so-called weakly symmetric Gelfand pairs (cf. [11, Example 4.3.2 and Exercise 4.3.3]); we develop the theory of spherical functions in an intrinsic way, that is, by regarding them as eigenfunctions of convolution operators (without using the decomposition of $\mathrm{Ind}_K^G\theta$ into irreducible representations) and obtain a characterization of spherical functions by means of a functional equation. This approach is suitable to more general settings, such as compact or locally compact groups: here we limit ourselves to the finite case since the main examples that we have discovered (and that we have fully analyzed) fall into this setting. Later (cf. Sect. 3.3), we express spherical functions as matrix coefficients of irreducible (spherical) representations. In Sect. 3.6, we prove a Frobenius–Schur type theorem for multiplicity-free triples (it provides a criterion for determining the type of a given irreducible spherical representation, namely being real, quaternionic, or complex).

As mentioned before, the case when θ is a one-dimensional representation and the example of the Gelfand–Graev representation of $\mathrm{GL}(2, \mathbb{F}_q)$ were developed, in full details, in [17, Chapters 13 and 14] (the last chapter is based on [53]; see also the pioneering work by Green [38]). Here (cf. Sect. 3.4) we recover the analysis of the one-dimensional case from the general theory we have developed so far and we briefly sketch the Gelfand–Graev example (cf. Sect. 3.5) in order to provide some of the necessary tools for our main new examples of multiplicity-free triples to which the second part of the monograph (Chaps. 5 and 6) is entirely devoted.

A particular case of interest is when the subgroup $K = N \leq G$ is normal (cf. Chap. 4). In the classical framework, (G, N) is a Gelfand pair if and only if the quotient group G/N is Abelian and, in this case, the spherical Fourier analysis simply reduces to the commutative harmonic analysis on G/N. In Chap. 4, we face the corresponding analysis for multiplicity-free triples of the form (G, N, θ), where θ is an irreducible N-representation. Now, G acts by conjugation on the dual of N and we denote by $I_G(\theta)$ the stabilizer of θ (this is the *inertia group* in Clifford theory; cf. the monographs by Berkovich and Zhmud [5], Huppert [43], and Isaacs [44]; see also [12, 14]). First of all, we study the commutant of $\mathrm{Ind}_N^G \theta$—we show that it is isomorphic to a modified convolution algebra on the quotient group $I_G(\theta)/N$—and we describe the associated Hecke algebra: all of this theory is developed without assuming multiplicity-freeness. We then prove that (G, N, θ) is a multiplicity-free triple if and only if $I_G(\theta)/N$ is Abelian and the multiplicity of θ in each irreducible representation of $I_G(\theta)$ is at most one. Moreover, if this is the case, the associated Hecke algebra is isomorphic to $L(I_G(\theta)/N)$, the (commutative) group algebra of $I_G(\theta)/N$ with its ordinary convolution product. Thus, as for Gelfand pairs, normality of the subgroup somehow trivializes the analysis of multiplicity-free triples.

As mentioned above, the last two chapters of the monographs are devoted to two examples of multiplicity-free triples constructed by means of the group $\mathrm{GL}(2, \mathbb{F}_q)$. Chapter 5 is devoted to the multiplicity-free triple $(\mathrm{GL}(2, \mathbb{F}_q), C, \nu_0)$, where C is the Cartan subgroup of $\mathrm{GL}(2, \mathbb{F}_q)$, which is isomorphic to the quadratic extension $\mathbb{F}_{q^2}$ of $\mathbb{F}_q$, and ν_0 is an *indecomposable* multiplicative character of $\mathbb{F}_{q^2}$ (that is, ν_0 is a character of the multiplicative group $\mathbb{F}_{q^2}^*$ such that $\nu_0(z)$ is *not* of the form $\psi(z\overline{z})$, $z \in \mathbb{F}_{q^2}^*$, where ψ is a multiplicative character of $\mathbb{F}_q$ and $\overline{z}$ is the conjugate of z). Actually, C is a *multiplicity-free subgroup*, that is, $(\mathrm{GL}(2, \mathbb{F}_q), C, \nu_0)$ is multiplicity-free for *every* multiplicative character ν_0. We remark that the case $\nu_0 = \iota_C$ (the trivial character of C) has been extensively studied by Terras under the name of *finite upper half plane* [69, Chapters 19, 20] and corresponds to the Gelfand pair $(\mathrm{GL}(2, \mathbb{F}_q), C)$. We have chosen to study, in full details, the indecomposable case because it is quite different from the Gelfand pair case analyzed by Terras and constitutes a new example, though much more difficult. We begin with a brief description of the representation theory of $\mathrm{GL}(2, \mathbb{F}_q)$, including the Kloosterman sums used for the cuspidal representations. We then compute the decomposition of $\mathrm{Ind}_C^{\mathrm{GL}(2,\mathbb{F}_q)} \nu_0$ into irreducible representations (cf. Sect. 5.4) and the corresponding spherical functions (cf. Sects. 5.5 and 5.6). We have developed new methods: in particular, in the study of the cuspidal representations, in order to circumvent some technical difficulties, we use, in a smart way, a projection formula onto a one-dimensional subspace.

In Chap. 6, we face the most important multiplicity-free triple of this monograph, namely $(\mathrm{GL}(2, \mathbb{F}_{q^2}), \mathrm{GL}(2, \mathbb{F}_q), \rho_\nu)$, where ρ_ν is a cuspidal representation. Now the representation that is induced is no more one-dimensional nor is itself an induced representation (as in the parabolic case). We have found an intriguing phenomenon: in the computations of the spherical functions associated with the corresponding parabolic spherical representations, we must use the results of Chap. 5, in particular the decomposition of an induced representation of the form $\mathrm{Ind}_C^{\mathrm{GL}(2,\mathbb{F}_q)}\xi$, with ξ a character of C. In other words, the methods developed in Chap. 5 (for the triple $(\mathrm{GL}(2,\mathbb{F}_q, C, \nu_0))$ turned out to be essential in the much more involved analysis of the second triple $(\mathrm{GL}(2,\mathbb{F}_{q^2}), \mathrm{GL}(2,\mathbb{F}_q), \rho_\nu)$.

We finally remark that it is not so difficult to find other examples of multiplicity-free induced representations within the framework of finite classical groups: for instance, as a consequence of the *branching rule* in the representation theory of the symmetric groups (see, e.g., [13]), S_n is a multiplicity-free subgroup of S_{n+1} for all $n \geq 1$. So, although the two examples that we have presented and fully analyzed here are new and highly nontrivial, we believe that we have only scraped the surface of the subject and that this deserves a wider investigation. For instance, in [3, 4] several Gelfand pairs constructed by means of $\mathrm{GL}(n,\mathbb{F}_q)$ and other finite linear groups are described. It would be interesting to analyze if some of these pairs give rise to multiplicity free triples by induction of a nontrivial representation.

We also mention that, very recently, a similar theory for locally compact groups has been developed by Ricci and Samanta in [56] (see also [47] and, for an earlier reference, the seminal paper by Godement [36]). In particular, their condition (0.1) (cf. also with (3) in [47]) corresponds exactly to our condition (2.1). In Sect. A.1, we show that the Gelfand–Graev representation yields a solution to a problem raised in the Introduction of their paper. For a classical account on the classification of Gelfand pairs on Lie groups, we refer to the monograph by Wolf [71] and the survey by Vinberg [70] (see also [72]).

The problem we addressed in the present monograph, namely the study of multiplicity-free induced representations, is certainly too general for a comprehensive and exhaustive study in its full generality. This is witnessed by the lack of literature (at our knowledge), somehow leading to believe that researchers were reluctant to attack such a general situation. So, although our research only barely scraped the surface of the subject, we believe that our monograph constitutes a serious attempt to obtain some reasonably meaningful new results in this direction.

We express our deep gratitude to Eiichi Bannai, Charles F. Dunkl, David Ginzburg, Pierre de la Harpe, Hiroshi Mizukawa, Akihiro Munemasa, Jean-Pierre Serre, Hajime Tanaka, Alain Valette, and the anonymous referees, for useful remarks and suggestions. Finally, we warmly thank Elena Griniari from Springer Verlag for her continuous encouragement and most precious help at various stages of the preparation of the manuscript.

Roma, Italy Tullio Ceccherini-Silberstein
Roma, Italy Fabio Scarabotti
Roma, Italy Filippo Tolli
May 2020

Contents

Symbols

Symbol	Definition	Page
$\langle \cdot, \cdot \rangle_V$	The scalar product on a vector space V	1
$\langle \cdot, \cdot \rangle_{\mathrm{Hom}(W,U)}$	The Hilbert–Schmidt scalar product on $\mathrm{Hom}(W, U)$	1
$\|\cdot\|_V$	The norm on a vector space V	1
$\|\cdot\|_{\mathrm{End}(W)}$	The Hilbert–Schmidt norm on $\mathrm{End}(W)$	2
δ_g	The Dirac function supported by $g \in G$	4
η	A generator of the (cyclic) group $\mathbb{F}_q^*$	49
$({}^g\theta, V)$	The g-conjugate of the irreducible N-representation (θ, V), $g \in G$, $N \trianglelefteq G$	56
(θ^s, V_s)	The K_s-representation associated with the K-representation (θ, V) and $s \in \mathcal{S}$	13
$\Theta(q)$	The unitary in $\mathrm{End}(V)$ conjugating $({}^g\theta, V)$ and (θ, V), $q \in Q$	57
$(\iota_G, \mathbb{C})$	The trivial representation of a group G	2
$(\lambda_G, L(G))$	The left regular representation of a group G	4
$(\lambda, L(G)^K)$	The permutation representation of G with respect to the subgroup K	4
μ	A multiplicative character of $\mathbb{F}_{q^2}$	77
μ	A generic indecomposable character of $\mathbb{F}_{q^4}$	95
ν	A multiplicative character of $\mathbb{F}_{q^2}$	73
ν	A fixed indecomposable multiplicative character of $\mathbb{F}_{q^2}$	95
ν_0	An indecomposable multiplicative character of $\mathbb{F}_{q^2}$	77
ξ	The $*$-isomorphism between $\widetilde{\mathscr{H}}(G, K, \theta)$ and $\mathrm{End}_G(\mathrm{Ind}_K^G V)$	15
ξ_1, ξ_2	Multiplicative characters of $\mathbb{F}_{q^2}$	95
Ξ	A right-inverse of the $*$-isomorphism ξ	18
$(\rho_G, L(G))$	The right regular representation of a group G	4
ρ_ν	The $(q-1)$-dimensional cuspidal representation of $\mathrm{GL}(2, \mathbb{F}_q)$, $\nu \in \widehat{\mathbb{F}_{q^2}^*}$	77
(σ, W)	A group representation $\sigma : G \to \mathrm{GL}(W)$ of a group G	2
(σ', W')	The conjugate of the group representation (σ, W)	50
$\sigma \oplus \rho$	The direct sum of two representations σ and ρ	3

Symbol	Definition	Page
$\sigma \preceq \rho$	The representation (σ, W) is contained in the representation (ρ, V)	3
τ	An antiautomortphism of a group G	32
τ	A unitary cocycle $H \times H \to \mathbb{T}$	54
τ_ρ	A unitary cocycle which is a coboundary $(\rho\colon H \to \mathbb{T})$	54
ϕ	A spherical function in the Hecke algebra $\mathscr{H}(G, K, \psi)$	35
ϕ^σ	The spherical function associated with $\sigma \in J$	44
ϕ^ν	The spherical function associated with the cuspidal representation ρ^ν	93
Φ	A linear functional on the Hecke algebra $\mathscr{H}(G, K, \psi)$	38
χ^ν	The character of the cuspidal representation ρ^ν	107
χ^σ	The character of the representation σ	6
χ_{ψ_1,ψ_2}	The one-dimensional representation of B associated with $\psi_1, \psi_2 \in \widehat{\mathbb{F}_q^*}$	75
$\widehat{\chi}_\psi^0$	The one-dimensional representation of $\mathrm{GL}(2, \mathbb{F}_q)$ associated with $\psi \in \widehat{\mathbb{F}_q^*}$	75
$\widehat{\chi}_\psi^1$	The q-dimensional parabolic representation of $\mathrm{GL}(2, \mathbb{F}_q)$, $\psi \in \widehat{\mathbb{F}_q^*}$	75
$\widehat{\chi}_{\psi_1,\psi_2}$	The $(q+1)$-dimensional parabolic representation of $\mathrm{GL}(2, \mathbb{F}_q)$, $\psi_1 \neq \psi_2 \in \widehat{\mathbb{F}_q^*}$	75
ψ	The element in $L(K)$ associated with (θ, V) and $v \in V$, $\\|v\\| = 1$	21
ψ	A multiplicative character of $\mathbb{F}_q$	77
Ψ	The decomposable character of $\mathbb{F}_{q^2}^*$ associated with $\psi \in \widehat{\mathbb{F}_q^*}$	76
$\mathrm{Aff}(\mathbb{F}_q)$	The affine group over $\mathbb{F}_q$	71
B	The Borel subgroup of $\mathrm{GL}(2, \mathbb{F}_q)$	48
B_1	The Borel subgroup of $\mathrm{GL}(2, \mathbb{F}_q)$	95
B_2	The Borel subgroup of $\mathrm{GL}(2, \mathbb{F}_{q^2})$	95
$\mathscr{B}(\mathbb{T}, H)$	The subgroup of all coboundaries of H	54
C	The Cartan subgroup of $\mathrm{GL}(2, \mathbb{F}_q)$	48
C_1	The Cartan subgroup of $\mathrm{GL}(2, \mathbb{F}_q)$	95
C_2	The Cartan subgroup of $\mathrm{GL}(2, \mathbb{F}_{q^2})$	95
$\mathscr{C}(\mathbb{T}, H)$	The group of all unitary cocycles $H \to \mathbb{T}$	54
d_σ	The dimension $\dim(W)$ of a group representation (σ, W)	2
D	The diagonal subgroup of $\mathrm{GL}(2, \mathbb{F}_q)$	49
E_ρ	The orthogonal projection onto the ρ-isotypic component of (σ, W)	7
E_σ	The orthogonal projection from $\mathscr{I}(G, K, \psi)$ onto $S_\sigma W_\sigma$	46
$\mathrm{End}(W)$	The algebra of all linear endomorphism $T\colon W \to W$	1
$\mathrm{End}_G(W)$	The commutant of (σ, W)	2
f^*	The function $f^*(g) = \overline{f(g^{-1})}$ for all $f \in L(G)$ and $g \in G$	4
$f_1 * f_2$	The convolution product of two functions $f_1, f_2 \in L(G)$	3
$f_1 *_\eta f_2$	The cocycle convolution product of $f_1, f_2 \in L(H)$ $(\eta \in \mathscr{C}(\mathbb{T}, H))$	55
F^*	The adjoint of an element $F \in \widetilde{\mathscr{H}}(G, K, \theta)$	12
$F_0(x, y)$	The function defined in (5.22)	84
$\widetilde{F_0}(x, y)$	The function defined in (5.27)	87
$F_1 * F_2$	The convolution product of elements $F_1, F_2 \in \widetilde{\mathscr{H}}(G, K, \theta)$	12

Symbol	Definition	Page
F_ν	The element in $V_{\widetilde{\chi}_{\psi_1,\psi_2}}$ associated with ν, $\nu^\sharp = \psi_1\psi_2$, $\psi_1 \neq \psi_2 \in \widehat{\mathbb{F}_q^*}$	79
$\mathscr{F}$	The spherical Fourier transform	45
$\mathbb{F}_q$	The field with $q := p^n$ elements	48
$\mathbb{F}_q^*$	The multiplicative subgroup of the field $\mathbb{F}_q$	49
$\mathbb{F}_{q^2}$	The quadratic extension of $\mathbb{F}_q$	49
$\widehat{G}$	The dual of a group G, i.e., a (fixed, once and for all) complete set of pairwise-inequivalent irreducible representations of G	2
$\mathscr{G}(G, H, K)$	The Greenhalgebra associated with $K \leq H \leq G$	125
$G_0(x, y)$	The function defined as $F_0(x, y)$ with μ_0 in place of ν_0	92
$\widetilde{G_0}(x, y)$	The function defined as $\widetilde{F}_{00}(x, y)$ with μ_0 in place of ν_0	92
G_1	The general linear group $\mathrm{GL}(2, \mathbb{F}_q)$	95
G_2	The general linear group $\mathrm{GL}(2, \mathbb{F}_{q^2})$	95
$\mathrm{GL}(W)$	The general linear group of the vector space W	1
$\mathrm{GL}(2, \mathbb{F}_q)$	The general linear group of rank 2 over $\mathbb{F}_q$	48
$\mathrm{Hom}(W, U)$	The space of all linear operators $T : W \to U$	1
$\mathrm{Hom}_G(W, U)$	The space of all intertwiners of the G-representations (σ, W) and (ρ, U)	2
$\widetilde{\mathscr{H}}(G, K, \theta)$	The Hecke algebra associated with the triple (G, K, θ)	11
$\mathscr{H}(G, K, \psi)$	The Hecke algebra associated with the triple (G, K, θ) and ψ	22
$\mathscr{H}^2(\mathbb{T}, H)$	The second cohomology group $\mathscr{C}(\mathbb{T}, H)/\mathscr{B}(\mathbb{T}, H)$ of H	54
$\pm i$	The square roots (in $\mathbb{F}_{q^2}$) of η	49
$I_G(\theta)$	The inertia group of the irreducible N-representation (θ, V), $N \trianglelefteq G$	57
I_V	The identity operator $V \to V$	42
$\mathscr{I}(G, K, \psi)$	The range $T_v(\mathrm{Ind}_K^G V) \subset L(G)$ of the operator T_v	22
$(\mathrm{Ind}_K^G \theta, \mathrm{Ind}_K^G V)$	The induction to the group $G \geq K$ of the K-representation (θ, V)	7
$j = j_{\chi,\nu}$	The generalized Kloosterman sum associated with nontrivial $\chi \in \widehat{\mathbb{F}_q}$ and indecomposable $\nu \in \widehat{\mathbb{F}_{q^2}^*}$	76
J	A complete set of pairwise-inequivalent irreducible G-representations contained in $\mathrm{Ind}_K^G \theta$	9
$J = J_{\widetilde{\chi},\mu}$	The generalized Kloosterman sum associated with nontrivial $\chi \in \widehat{\mathbb{F}_q}$, cf. (6.19), and indecomposable $\mu \in \widehat{\mathbb{F}_{q^4}^*}$	104
K_s	The subgroup $K \cap sKs^{-1}$, where $s \in \mathscr{S}$	12
L	The isometric immersion $L(\mathbb{F}_q^*) \to L(\mathbb{F}_{q^2}^*)$	107
$L(H)_\eta$	The twisted group algebra of H, $\eta \in \mathscr{C}(\mathbb{T}, H)$	55
$L(G)$	The group algebra of G	3
$L(G)^K$	The $L(G)$-subalgebra of K-right-invariant functions on G	4
$L(G, H, K)$	The Hirschman subalgebra of $H - K$-invariant functions in $L(G)$	125
L_θ	The operator $L(\mathbb{F}_q^*) \to L(\mathbb{F}_{q^2}^*)$, $\theta \in \mathbb{F}_q$	105
L_σ	The isometric map $V \to W$ such that $\mathrm{Hom}_K(V, \mathrm{Res}_K^G W) = \mathbb{C}L_\sigma$	41
$\mathscr{L}_T$	The element in $\widetilde{\mathscr{H}}(G, K, \theta)$ associated with $T \in \mathrm{Hom}_{K_s}(\mathrm{Res}_{K_s}^K \theta, \theta^s)$	13
${}^K L(G)^K$	The $L(G)$-subalgebra of bi-K-invariant functions on G	4

Symbol	Definition	Page
m_σ^ρ	The multiplicity of the representation σ in the representation ρ	3
P	The orthogonal projection of $L(G)$ onto $\mathscr{I}(G, K, \psi)$	22
P	The projection of $L(\mathbb{F}^*_{q^2})$ onto the subspace of ρ_ν	107
Q	A complete set of representatives for the cosets of N in $I_G(\theta)$, $1_G \in Q$	56
Q_1	The operator $L(\mathbb{F}^*_{q^2}) \to L(\mathbb{F}^*_{q^2})$	107
$(\mathrm{Res}^G_K \sigma, W)$	The restriction of the G-representation (σ, W) to the subgroup $K \leq G$	7
$\mathscr{S}$	A complete set of representatives for the double K-cosets in G	12
$\mathscr{S}_0$	The set of all $s \in \mathscr{S}$ such that $\mathrm{Hom}_{K_s}(\mathrm{Res}^K_{K_s} \theta, \theta^s)$ is nontrivial	14
$\mathscr{S}(G, K, \psi)$	The set of all spherical functions in the Hecke algebra $\mathscr{H}(G, K, \psi)$	35
S_σ	The isometric immersion of W_σ into $\mathscr{I}(G, K, \psi)$	43
S_v	The $*$-antiisomorphism from $\widetilde{\mathscr{H}}(G, K, \theta)$ onto $\mathscr{H}(G, K, \psi)$	22
$\mathrm{tr}(\cdot)$	The trace of linear operators	1
$T \mapsto T^\sharp$	An antiautomorphism of the algebra $\mathrm{End}(V)$	32
$\mathscr{T}$	A transversal for the left-cosets of K in G	8
T_σ	The isometric immersion of W into $\mathrm{Ind}^G_K V$	41
T_v	The isometry in $\mathrm{Hom}_G(\mathrm{Ind}^G_K V, L(G))$	22
$\mathbb{T}$	The group $\mathbb{R}/\mathbb{Z}$	54
T^*	The adjoint $\in \mathrm{Hom}(U, W)$ of the linear operators $T \in \mathrm{Hom}(W, U)$	1
T_f	The convolution operator in $\mathrm{End}_G(L(G))$ with kernel $f \in L(G)$	5
$u^\sigma_{i,j}$	A matrix coefficient for the representation σ	6
U	The unipotent subgroup of $\mathrm{GL}(2, \mathbb{F}_q)$	49
U_1	The unipotent subgroup of $\mathrm{GL}(2, \mathbb{F}_q)$	95
U_2	The unipotent subgroup of $\mathrm{GL}(2, \mathbb{F}_{q^2})$	95
U_ϕ	The spherical representation associated with the spherical function ϕ	40
w	The matrix $\begin{pmatrix} 0 & 1 \\ 1 & 0 \end{pmatrix} \in \mathrm{GL}(2, \mathbb{F}_q)$	49
w^σ	The vector $L_\sigma v \in W_\sigma, \sigma \in J$	43
W	The matrix $\begin{pmatrix} i & 1 \\ 1 & 0 \end{pmatrix} \in \mathrm{GL}(2, \mathbb{F}_{q^2})$	96
W^K	The subspace of K-invariant vectors of W	4
W_σ	The representation space of the spherical representation $\sigma \in J$	43
$W^\perp$	The orthogonal complement of the subspace $W \subset U$	3
Z	The center of $\mathrm{GL}(2, \mathbb{F}_q)$	49

Chapter 1
Preliminaries

In this chapter, we fix notation and recall some basic facts on linear algebra and representation theory of finite groups that will be used in the proofs of several results in the sequel.

1.1 Representations of Finite Groups

All vector spaces considered here are complex. Moreover, we shall equip every finite dimensional vector space V with a scalar product denoted by $\langle \cdot, \cdot \rangle_V$ and associated norm $\|\cdot\|_V$; we usually omit the subscript if the vector space we are referring to is clear from the context. Given two finite dimensional vector spaces W and U, we denote by $\mathrm{Hom}(W, U)$ the vector space of all linear maps from W to U. When $U = W$ we write $\mathrm{End}(W) = \mathrm{Hom}(W, W)$ and denote by $\mathrm{GL}(W) \subseteq \mathrm{End}(W)$ the general linear group of W consisting of all bijective linear self-maps of W. Also, for $T \in \mathrm{Hom}(W, U)$ we denote by $T^* \in \mathrm{Hom}(U, W)$ the adjoint of T.

We define a (normalized) *Hilbert-Schmidt* scalar product on $\mathrm{Hom}(W, U)$ by setting

$$\langle T_1, T_2 \rangle_{\mathrm{Hom}(W,U)} = \frac{1}{\dim W} \mathrm{tr}(T_2^* T_1) \tag{1.1}$$

for all $T_1, T_2 \in \mathrm{Hom}(W, U)$, where $\mathrm{tr}(\cdot)$ denotes the trace of linear operators; note that this scalar product (as well as all other scalar products which we shall introduce thereafter) is conjugate-linear in the *second* argument. Moreover, by centrality of the trace (so that $\mathrm{tr}(T_2^* T_1) = \mathrm{tr}(T_1 T_2^*)$), we have

$$\langle T_1, T_2 \rangle_{\mathrm{Hom}(W,U)} = \frac{\dim U}{\dim W} \langle T_2^*, T_1^* \rangle_{\mathrm{Hom}(U,W)}. \tag{1.2}$$

T. Ceccherini-Silberstein et al., *Gelfand Triples and Their Hecke Algebras*,
Lecture Notes in Mathematics 2267, https://doi.org/10.1007/978-3-030-51607-9_1

In particular, the map $T \mapsto \sqrt{\dim U / \dim W} T^*$ is an isometry from $\mathrm{Hom}(W, U)$ onto $\mathrm{Hom}(U, W)$. Finally, note that denoting by $I_W : W \to W$ the identity operator, we have $\|I_W\|_{\mathrm{End}(W)} = 1$.

We now recall some basic facts on the representation theory of finite groups. For more details we refer to our monographs [11, 13, 17]. Let G be a finite group. A *unitary representation* of G is a pair (σ, W) where W is a finite dimensional vector space and $\sigma : G \to \mathrm{GL}(W)$ is a group homomorphism such that $\sigma(g)$ is unitary (that is, $\sigma(g)^*\sigma(g) = I_W$) for all $g \in G$. In the sequel, the term "unitary" will be omitted. We denote by $d_\sigma = \dim(W)$ the dimension of the representation (σ, W). We denote by $(\iota_G, \mathbb{C})$ the *trivial representation* of G, that is, the one-dimensional G-representation defined by $\iota_G(g) = \mathrm{Id}_{\mathbb{C}}$ for all $g \in G$.

Let (σ, W) be a G-representation. A subspace $V \leq W$ is said to be *G-invariant* provided $\sigma(g)V \subseteq V$ for all $g \in G$. Writing $\sigma|_V(g) = \sigma(g)|_V$ for all $g \in G$, we have that $(\sigma|_V, V)$ is a G-representation, called a *subrepresentation* of σ. We then write $\sigma|_V \leq \sigma$. One says that σ is *irreducible* provided the only G-invariant subspaces are trivial (equivalently, σ admits no proper subrepresentations).

Let (σ, W) and (ρ, U) be two G-representations. We denote by

$$\mathrm{Hom}_G(W, U) = \{T \in \mathrm{Hom}(W, U) : T\sigma(g) = \rho(g)T, \text{ for all } g \in G\},$$

the space of all *intertwining operators*. When $U = W$ we write $\mathrm{End}_G(W) = \mathrm{Hom}_G(W, W)$. We equip $\mathrm{Hom}_G(W, U)$ with a scalar product by restricting the Hilbert-Schmidt scalar product (1.1).

Observe that if $T \in \mathrm{Hom}_G(W, U)$ then $T^* \in \mathrm{Hom}_G(U, W)$. Indeed, for all $g \in G$,

$$T^*\rho(g) = T^*\rho(g^{-1})^* = (\rho(g^{-1})T)^* = (T\sigma(g^{-1}))^* = \sigma(g^{-1})^*T^* = \sigma(g)T^*. \tag{1.3}$$

One says that (σ, W) and (ρ, U) are *equivalent*, and we shall write $(\sigma, W) \sim (\rho, U)$ (or simply $\sigma \sim \rho$), if there exists a bijective intertwining operator $T \in \mathrm{Hom}_G(W, U)$.

The vector space $\mathrm{End}_G(W)$ of all intertwining operators of (σ, W) with itself, when equipped with the multiplication given by the composition of maps and the adjoint operation is a $*$-algebra (see [13, Chapter 7], [17, Sections 10.3 and 10.6]), called the *commutant* of (σ, W). We can thus express the well known *Schur's lemma* as follows: (σ, W) is irreducible if and only if its commutant is one-dimensional (as a vector space), that is, it reduces to the scalars (the scalar multiples of the identity I_W).

We denote by $\widehat{G}$ a (fixed, once and for all) complete set of pairwise-inequivalent irreducible representations of G and we refer to it as to the *dual* of G. It is well known (cf. [11, Theorem 3.9.10] or [17, Theorem 10.3.13.(ii)]) that the cardinality of $\widehat{G}$ equals the number of conjugacy classes in G so that, in particular, $\widehat{G}$ is finite. Moreover, if $\sigma, \rho \in \widehat{G}$ we set $\delta_{\sigma,\rho} = 1$ (resp. $= 0$) if $\sigma = \rho$ (resp. otherwise).

Let (σ, W) and (ρ, U) be two G-representations.

The *direct sum* of σ and ρ is the representation $(\sigma \oplus \rho, W \oplus U)$ defined by $[(\sigma \oplus \rho)(g)](w, u) = (\sigma(g)w, \rho(g)u)$ for all $g \in G$, $w \in W$ and $u \in U$.

Moreover, if σ is a subrepresentation of ρ, then denoting by $W^\perp = \{u \in U : \langle u, w\rangle_U = 0 \text{ for all } w \in W\}$ the orthogonal complement of W in U, we have that $W^\perp$ is a G-invariant subspace and $\rho = \sigma \oplus \rho|_{W^\perp}$. From this, one deduces that every representation ρ decomposes as a (finite) direct sum of irreducible subrepresentations. More generally, when σ is equivalent to a subrepresentation of ρ, we say that σ is *contained* in ρ and we write $\sigma \preceq \rho$ (clearly, if $\sigma \leq \rho$ then $\sigma \preceq \rho$).

Suppose that (σ, W) is irreducible. Then the number $m = m_\sigma^\rho = \dim \mathrm{Hom}_G(W, U)$ denotes the *multiplicity* of σ in ρ. This means that one may decompose $U = U_1 \oplus U_2 \oplus \cdots \oplus U_m \oplus U_{m+1}$ with $(\rho|_{U_i}, U_i) \sim (\sigma, W)$ for all $i = 1, 2, \ldots, m$ and σ is not contained in $\rho|_{U_{m+1}}$. The G-invariant subspace $U_1 \oplus U_2 \oplus \cdots \oplus U_m \leq U$ is called the W-*isotypic component* of U and is denoted by mW. One also says that ρ (or, equivalently, U) contains m copies of σ (resp. of W). If this is the case, we say that $T_1, T_2, \ldots, T_m \in \mathrm{Hom}_G(W, U)$ yield an *isometric orthogonal decomposition* of mW if $T_i \in \mathrm{Hom}_G(W, U)$, $T_i W \leq U \ominus U_{m+1}$, and, in addition,

$$\langle T_i w_1, T_j w_2\rangle_U = \langle w_1, w_2\rangle_W \delta_{i,j} \tag{1.4}$$

for all $w_1, w_2 \in W$ and $i, j = 1, 2, \ldots, m$. This implies that the subrepresentation $mW = U_1 \oplus U_2 \oplus \cdots \oplus U_m$ is equal to the *orthogonal* direct sum $T_1 W \oplus T_2 W \oplus \cdots \oplus T_m W$, and each operator T_j is a isometry from W onto $U_j \equiv TW_j$. For a quite detailed analysis of this decomposition, we refer to [17, Section 10.6].

Finally, a representation (ρ, U) is *multiplicity-free* if every $(\sigma, W) \in \widehat{G}$ has multiplicity at most one in ρ, that is, $\dim \mathrm{Hom}_G(W, U) \leq 1$. In other words, given a decomposition of $\rho = \rho_1 \oplus \rho_2 \oplus \cdots \oplus \rho_n$ into irreducible subrepresentations, the ρ_i's are pairwise inequivalent. Alternatively, as suggested by de la Harpe [40], one has that (ρ, U) is multiplicity-free if for any nontrivial decomposition $\rho = \rho_1 \oplus \rho_2$ (with (ρ_1, U_1) and (ρ_2, U_2) not necessarily irreducible) there is no $(\sigma, W) \in \widehat{G}$ such that $\sigma \preceq \rho_i$ (i.e., $\dim \mathrm{Hom}_G(W, U_i) \geq 1$) for $i = 1, 2$. The equivalence between the two definitions is an immediate consequence of the isomorphism $\mathrm{Hom}_G(W, U_1 \oplus U_2) \cong \mathrm{Hom}_G(W, U_1) \oplus \mathrm{Hom}_G(W, U_1)$.

1.2 The Group Algebra, the Left-Regular and the Permutation Representations, and Gelfand Pairs

We denote by $L(G)$ the group algebra of G.

This is the vector space of all functions $f \colon G \to \mathbb{C}$ equipped with the *convolution product* $*$ defined by setting $[f_1 * f_2](g) = \sum_{h \in G} f_1(h) f_2(h^{-1}g) = \sum_{h \in G} f_1(gh) f_2(h^{-1})$, for all $f_1, f_2 \in L(G)$ and $g \in G$. We shall endow $L(G)$

with the scalar product $\langle \cdot, \cdot \rangle_{L(G)}$ defined by setting

$$\langle f_1, f_2 \rangle_{L(G)} = \sum_{g \in G} f_1(g) \overline{f_2(g)} \tag{1.5}$$

for all $f_1, f_2 \in L(G)$. The *Dirac functions* δ_g, defined by $\delta_g(g) = 1$ and $\delta_g(h) = 0$ if $h \neq g$, for all $g, h \in G$, constitute a natural orthonormal basis for $L(G)$. We shall also equip $L(G)$ with the *involution* $f \mapsto f^*$, where $f^*(g) = \overline{f(g^{-1})}$, for all $f \in L(G)$ and $g \in G$. It is straightforward to check that $(f_1 * f_2)^* = f_2^* * f_1^*$, for all $f_1, f_2 \in L(G)$. We shall thus regard $L(G)$ as a $*$-*algebra.*

The *left-regular representation* of G is the G-representation $(\lambda_G, L(G))$ defined by setting $[\lambda_G(h)f](g) = f(h^{-1}g)$, for all $f \in L(G)$ and $h, g \in G$. Similarly, the *right-regular representation* of G is the G-representation $(\rho_G, L(G))$ defined by setting $[\rho_G(h)f](g) = f(gh)$, for all $f \in L(G)$ and $h, g \in G$. Note that the left-regular and right-regular representations commute, that is,

$$\lambda_G(g_1)\rho_G(g_2) = \rho_G(g_2)\lambda_G(g_1) \tag{1.6}$$

for all $g_1, g_2 \in G$.

Given a subgroup $K \leq G$ we denote by

$$L(G)^K = \{f \in L(G) : f(gk) = f(g), \text{ for all } g \in G, k \in K\}$$

and

$${}^K L(G)^K = \{f \in L(G) : f(k_1 g k_2) = f(g), \text{ for all } g \in G, k_1, k_2 \in K\}$$

the $L(G)$-subalgebras of K-*right-invariant* and *bi*-K-*invariant* functions on G, respectively. Note that the subspace $L(G)^K \leq L(G)$ is G-invariant with respect to the left-regular representation. The G-representation $(\lambda, L(G)^K)$, where $\lambda(g)f = \lambda_G(g)f$ for all $g \in G$ and $f \in L(G)^K$ (equivalently, $\lambda = \lambda_G|_{L(G)^K}$) is called the *permutation representation* of G with respect to the subgroup K.

More generally, given a representation (σ, W) we denote by

$$W^K = \{w \in W : \sigma(k)w = w, \text{ for all } k \in K\} \leq W$$

the subspace of K-*invariant vectors* of W. This way, if $(\sigma, W) = (\rho_G, L(G))$ we have $(L(G))^K = L(G)^K$ while, if $(\sigma, W) = (\lambda, L(G)^K)$ we have $\left(L(G)^K\right)^K = {}^K L(G)^K$.

For the following result we refer to [10] and/or to the monographs [11, Chapter 4] and [24].

Theorem 1.1 *The following conditions are equivalent:*

(a) *The algebra* ${}^K L(G)^K$ *is commutative;*
(b) *the permutation representation* $(\lambda, L(G)^K)$ *is multiplicity-free;*
(c) *the algebra* $\mathrm{End}_G(L(G)^K)$ *is commutative;*
(d) *for every* $(\sigma, W) \in \widehat{G}$ *one has* $\dim(W^K) \leq 1$*;*
(e) *for every* $(\sigma, W) \in \widehat{G}$ *one has* $\dim \mathrm{Hom}_G(W, L(G)^K) \leq 1$.

Note that the equivalence (a) $\Leftrightarrow$ (c) follows from the anti-isomorphism (1.11) below.

Definition 1.1 If one of the equivalent conditions in Theorem 1.1 is satisfied, one says that (G, K) is a *Gelfand pair.*

1.3 The Commutant of the Left-Regular and Permutation Representations

Given $f \in L(G)$, the (right) *convolution operator* with *kernel* f is the linear map $T_f : L(G) \to L(G)$ defined by

$$T_f f' = f' * f \tag{1.7}$$

for all $f' \in L(G)$. We have

$$T_{f_1 * f_2} = T_{f_2} T_{f_1} \tag{1.8}$$

and

$$T_{f^*} = (T_f)^*, \tag{1.9}$$

for all f_1, f_2 and f in $L(G)$. Moreover, $T_f \in \mathrm{End}_G(L(G))$ (this is a consequence of (1.6)) and the map

$$\begin{array}{rcl} L(G) & \longrightarrow & \mathrm{End}_G(L(G)) \\ f & \longmapsto & T_f \end{array} \tag{1.10}$$

is a $*$-anti-isomorphism of $*$-algebras (see [13, Proposition 1.5.2] or [17, Proposition 10.3.5]). Note that the restriction of the map (1.10) to the subalgebra ${}^K L(G)^K$ of bi-K-invariant functions on G yields a $*$-anti-isomorphism

$${}^K L(G)^K \to \mathrm{End}_G(L(G)^K). \tag{1.11}$$

It is easy to check that $T_f\delta_g = \lambda_G(g)f$ and $\mathrm{tr}(T_f) = |G|f(1_G)$. We deduce that

$$\begin{aligned}\langle T_{f_1}, T_{f_2}\rangle_{\mathrm{End}(L(G))} &= \frac{1}{|G|}\mathrm{tr}\left[(T_{f_2})^* T_{f_1}\right] = \frac{1}{|G|}\mathrm{tr}\left[T_{f_1 * f_2^*}\right] \\ &= \left[f_1 * f_2^*\right](1_G) = \langle f_1, f_2\rangle_{L(G)} \end{aligned} \tag{1.12}$$

for all $f_1, f_2 \in L(G)$. This shows that the map (1.10) is an isometry.

Let (σ, W) be a representation of G and let $\{w_1, w_2, \ldots, w_{d_\sigma}\}$ be an orthonormal basis of W. The corresponding *matrix coefficients* $u^\sigma_{j,i} \in L(G)$ are defined by setting

$$u^\sigma_{j,i}(g) = \langle \sigma(g)w_i, w_j\rangle \tag{1.13}$$

for all $i, j = 1, 2, \ldots, d_\sigma$ and $g \in G$.

Proposition 1.1 *Let $\sigma, \rho \in \widehat{G}$. Then*

$$\langle u^\sigma_{i,j}, u^\rho_{h,k}\rangle = \frac{|G|}{d_\sigma}\delta_{\sigma,\rho}\delta_{i,h}\delta_{j,k} \quad \text{(orthogonality relations)}, \tag{1.14}$$

$$u^\sigma_{i,j} * u^\rho_{h,k} = \frac{|G|}{d_\sigma}\delta_{\sigma,\rho}\delta_{j,h}u^\sigma_{i,k} \quad \text{(convolution properties)}, \tag{1.15}$$

and

$$u^\sigma_{i,j}(g_1g_2) = \sum_{\ell=1}^{d_\sigma} u^\sigma_{i,\ell}(g_1)u^\sigma_{\ell,j}(g_2) \tag{1.16}$$

for all $i, j = 1, 2, \ldots, d_\sigma$, $h, k = 1, 2, \ldots, d_\rho$, and $g_1, g_2 \in G$.

Proof See [11, Lemma 3.6.3 and Lemma 3.9.14] or [17, Lemma 10.2.10, Lemma 10.2.13, and Proposition 10.3.6]. □

The sum $\chi^\sigma = \sum_{i=1}^{d_\sigma} u^\sigma_{i,i} \in L(G)$ of the diagonal entries of the matrix coefficients is called the *character* of σ. Note that $\chi^\sigma(g) = \mathrm{tr}(\sigma(g))$ for all $g \in G$. The following elementary formula is a generalization of [11, Exercise 9.5.8.(2)] (see also [17, Proposition 10.2.26)]).

Proposition 1.2 *Suppose (σ, W) is irreducible and let $w \in W$ be a vector of norm 1. Consider the associated diagonal matrix coefficient $\phi_w \in L(G)$ defined by $\phi_w(g) = \langle\sigma(g)w, w\rangle$ for all $g \in G$. Then*

$$\chi^\sigma(g) = \frac{d_\sigma}{|G|}\sum_{h\in G}\phi_w(h^{-1}gh) \tag{1.17}$$

for all $g \in G$.

Proof Let $\{w_1, w_2, \ldots, w_{d_\sigma}\}$ be an orthonormal basis of W with $w_1 = w$ and let $u^\sigma_{j,i}$ as in (1.13). Then, for all $g \in G$ and $i = 1, 2, \ldots, d_\sigma$, we have

$$\sigma(g)w_i = \sum_{j=1}^{d_\sigma} u^\sigma_{j,i}(g) w_j$$

so that

$$\begin{aligned}
\sum_{h\in G} \phi_w(h^{-1}gh) &= \sum_{h\in G} \langle \sigma(g)\sigma(h)w_1, \sigma(h)w_1\rangle \\
&= \sum_{j,\ell=1}^{d_\sigma} \sum_{h\in G} u^\sigma_{j,1}(h)\overline{u^\sigma_{\ell,1}(h)} \langle \sigma(g)w_j, w_\ell\rangle \\
&= \sum_{j,\ell=1}^{d_\sigma} \langle u^\sigma_{j,1} u^\sigma_{\ell,1}\rangle \langle \sigma(g)w_j, w_\ell\rangle \\
\text{(by (1.14))} &= \frac{|G|}{d_\sigma} \chi^\sigma(g),
\end{aligned}$$

and (1.17) follows. □

From [13, Corollary 1.3.15] we recall the following fact. Let (σ, W) and (ρ, V) be two G-representations and suppose that ρ is irreducible and contained in σ. Then

$$E_\rho = \frac{d_\rho}{|G|} \sum_{g\in G} \overline{\chi^\rho(g)} \sigma(g) \tag{1.18}$$

is the orthogonal projection onto the ρ-isotypic component of W.

1.4 Induced Representations

Let now $K \leq G$ be a subgroup. We denote by $(\mathrm{Res}^G_K \sigma, W)$ the *restriction* of the G-representation (σ, W) to K, that is, the K-representation defined by $[\mathrm{Res}^G_K \sigma](k) = \sigma(k)$ for all $k \in K$.

Given a K-representation (θ, V) of K, denote by $\lambda = \mathrm{Ind}^G_K \theta$ the *induced representation* (see, for instance, [7, 9, 13, 14, 17, 32, 52, 65, 66, 69]). We recall that the representation space of λ is given by

$$\mathrm{Ind}^G_K V = \{f : G \to V \text{such that } f(gk) = \theta(k^{-1}) f(g), \text{ for all } g \in G, k \in K\} \tag{1.19}$$

and that

$$[\lambda(g)f](g') = f(g^{-1}g'), \tag{1.20}$$

for all $f \in \mathrm{Ind}_K^G V$ and $g, g' \in G$.

As an example, one checks that if $(\iota_K, \mathbb{C})$ is the trivial representation of K, then $(\mathrm{Ind}_K^G \iota_K, \mathrm{Ind}_K^G \mathbb{C})$ equals the permutation representation $(\lambda, L(G)^K)$ of G with respect to the subgroup K (see [14, Proposition 1.1.7] or [17, Example 11.1.6]).

Let $\mathscr{T} \subseteq G$ be a left-transversal for K, that is, a complete set of representatives for the left-cosets gK of K in G. Then we have the decomposition

$$G = \bigsqcup_{t \in \mathscr{T}} tK, \tag{1.21}$$

where, from now on, $\bigsqcup$ denotes a disjoint union. For $v \in V$ we define $f_v \in \mathrm{Ind}_K^G V$ by setting

$$f_v(g) = \begin{cases} \theta(g^{-1})v & \text{if } g \in K \\ 0 & \text{otherwise.} \end{cases} \tag{1.22}$$

Then, for every $f \in \mathrm{Ind}_K^G V$, we have

$$f = \sum_{t \in \mathscr{T}} \lambda(t) f_{v_t} \tag{1.23}$$

where $v_t = f(t)$ for all $t \in \mathscr{T}$. The induced representation $\mathrm{Ind}_K^G \theta$ is unitary with respect to the scalar product $\langle \cdot, \cdot \rangle_{\mathrm{Ind}_K^G V}$ defined by

$$\langle f_1, f_2 \rangle_{\mathrm{Ind}_K^G V} = \frac{1}{|K|} \sum_{g \in G} \langle f_1(g), f_2(g) \rangle_V = \sum_{t \in \mathscr{T}} \langle f_1(t), f_2(t) \rangle_V \tag{1.24}$$

for all $f_1, f_2 \in \mathrm{Ind}_K^G V$. Moreover, if $\{v_j : j = 1, 2, \ldots, d_\theta\}$ is an orthonormal basis in V then the set

$$\{\lambda(t) f_{v_j} : t \in \mathscr{T}, j = 1, 2, \ldots, d_\theta\} \tag{1.25}$$

is an orthonormal basis in $\mathrm{Ind}_K^G V$ (see [9, Theorem 2.1] and [17, Theorem 11.1.11]).

A well known relation between the induction of a K-representation (θ, V) and a G-representation (σ, W) is expressed by the so called *Frobenius reciprocity* (cf. [13, Theorem 1.6.11], [14, Theorem 1.1.19], or [17, Theorem 11.2.1]):

$$\mathrm{Hom}_G(W, \mathrm{Ind}_K^G V) \cong \mathrm{Hom}_K(\mathrm{Res}_G^K W, V). \tag{1.26}$$

Let $J = \{\sigma \in \widehat{G} : \sigma \preceq \mathrm{Ind}_K^G \theta\}$ denote a complete set of pairwise inequivalent irreducible G-representations contained in $\mathrm{Ind}_K^G \theta$. For $\sigma \in J$ we denote by W_σ its representation space and by $m_\sigma = \dim \mathrm{Hom}_G(\sigma, \mathrm{Ind}_K^G \theta) \geq 1$ its multiplicity in $\mathrm{Ind}_K^G \theta$. Then

$$\mathrm{Ind}_K^G V \cong \bigoplus_{\sigma \in J} m_\sigma W_\sigma \tag{1.27}$$

is the decomposition of $\mathrm{Ind}_K^G V$ into irreducible G-representations and we have the $*$-isomorphism of $*$-algebras

$$\mathrm{End}_G(\mathrm{Ind}_K^G V) \cong \bigoplus_{\sigma \in J} M_{m_\sigma}(\mathbb{C}), \tag{1.28}$$

where $M_m(\mathbb{C})$ denotes the $*$-algebra of all $m \times m$ complex matrices (cf. [17, Theorem 10.6.3]). In particular:

Proposition 1.3 *The following conditions are equivalent:*

(a) $\mathrm{Ind}_K^G \theta$ *is multiplicity-free (that is,* $m_\sigma = 1$ *for all* $\sigma \in J$*);*
(b) *the algebra* $\mathrm{End}_G(\mathrm{Ind}_K^G V)$ *(i.e. the commutant of* $\mathrm{Ind}_G^K V$*) is commutative;*
(c) $\mathrm{End}_G(\mathrm{Ind}_K^G V)$ *is isomorphic to the* $*$*-algebra* $\mathbb{C}^J = \{f : J \to \mathbb{C}\}$ *equipped with pointwise multiplication and complex conjugation.*

Remark 1.1 In (1.27) and (1.28) we have used the symbol $\cong$ to denote an *isomorphism* (with respect to the corresponding algebraic structure). We will use the equality symbol $=$ to denote an *explicit decomposition*. For instance, in the multiplicity-free case this corresponds to a choice of an *isometric immersion* of W_σ into $\mathrm{Ind}_K^G V$, that is, to a map $T_\sigma \in \mathrm{Hom}_G(W_\sigma, \mathrm{Ind}_K^G V)$ which is also an isometry. Clearly, in this case, $\mathrm{Hom}_G(W_\sigma, \mathrm{Ind}_K^G V) = \mathbb{C}T_\sigma$ and

$$\mathrm{Ind}_K^G V = \bigoplus_{\sigma \in J} T_\sigma W_\sigma \tag{1.29}$$

is the explicit decomposition. If multiplicities arise, then we decompose explicitly each isotypic component as in Sect. 1.1 (cf. (1.4)).

Chapter 2
Hecke Algebras

Let G be a finite group and $K \leq G$ a subgroup. Recalling the equality between the induced representation $(\mathrm{Ind}_K^G \iota_K, \mathrm{Ind}_K^G \mathbb{C})$ and the permutation representation $(\lambda, L(G)^K)$, (1.11) yields a $*$-algebra isomorphism between the algebra of bi-K-invariant functions on G and the commutant of the representation obtained by inducing to G the trivial representation of K.

In Sect. 2.1, expanding the ideas in [7, Theorem 34.1] and in [53, Section 3], we generalize this fact by showing that for a generic representation (θ, V) of K, the commutant of $\mathrm{Ind}_G^K V$, that is, $\mathrm{End}_G(\mathrm{Ind}_K^G V)$, is isomorphic to a suitable convolution algebra of operator-valued maps on G. This may be considered as a detailed formulation of Mackey's formula for invariants (see [9, Section 6] or [17, Corollary 11.4.4]).

Later, in Sect. 2.2, we show that, when θ is irreducible, the algebra $\mathrm{End}_G(\mathrm{Ind}_K^G V)$ is isomorphic to a suitable subalgebra of the group algebra $L(G)$ of G.

2.1 Mackey's Formula for Invariants Revisited

In this section, we study isomorphisms (or antiisomorphisms) between three $*$-algebras. We explicitly use the terminology of a Hecke algebra only for the first one (cf. Definition 2.1), although one may carry it also for the other two.

Let (θ, V) be a K-representation. We denote by $\widetilde{\mathscr{H}}(G, K, \theta)$ the set of all maps $F \colon G \to \mathrm{End}(V)$ such that

$$F(k_1 g k_2) = \theta(k_2^{-1}) F(g) \theta(k_1^{-1}), \text{ for all } g \in G \text{ and } k_1, k_2 \in K. \tag{2.1}$$

T. Ceccherini-Silberstein et al., *Gelfand Triples and Their Hecke Algebras*,
Lecture Notes in Mathematics 2267, https://doi.org/10.1007/978-3-030-51607-9_2

Given $F_1, F_2 \in \widetilde{\mathscr{H}}(G, K, \theta)$ we define their *convolution product* $F_1 * F_2 \colon G \to \mathrm{End}(V)$ by setting

$$[F_1 * F_2](g) = \sum_{h \in G} F_1(h^{-1}g)F_2(h) \tag{2.2}$$

for all $g \in G$, and their scalar product as

$$\langle F_1, F_2 \rangle_{\widetilde{\mathscr{H}}(G,K,\theta)} = \sum_{g \in G} \langle F_1(g), F_2(g) \rangle_{\mathrm{End}(V)}. \tag{2.3}$$

Finally, for $F \in \widetilde{\mathscr{H}}(G, K, \theta)$ we define the *adjoint* $F^* \colon G \to \mathrm{End}(V)$ by setting

$$F^*(g) = [F(g^{-1})]^* \tag{2.4}$$

for all $g \in G$, where $[F(g^{-1})]^*$ is the adjoint of the operator $F(g^{-1}) \in \mathrm{End}(V)$.

It is easy to check that $\widetilde{\mathscr{H}}(G, K, \theta)$ is an associative unital algebra with respect to this convolution. The identity is the function F_0 defined by setting $F_0(k) = \frac{1}{|K|}\theta(k^{-1})$ for all $k \in K$ and $F_0(g) = 0$ for $g \in G$ not in K; see also (2.9) below. Moreover, F^* still belongs to $\widetilde{\mathscr{H}}(G, K, \theta)$, the map $F \mapsto F^*$ is an involution, that is, $(F^*)^* = F$, and $(F_1 * F_2)^* = F_2^* * F_1^*$, for all $F, F_1, F_2 \in \widetilde{\mathscr{H}}(G, K, \theta)$.

Definition 2.1 The unital $*$-algebra $\widetilde{\mathscr{H}}(G, K, \theta)$ is called the *Hecke algebra* associated with the group G and the K-representation (θ, V).

Let $\mathscr{S} \subseteq G$ be a complete set of representatives for the double K-cosets in G so that

$$G = \bigsqcup_{s \in \mathscr{S}} KsK. \tag{2.5}$$

We assume that $1_G \in \mathscr{S}$, that is, 1_G is the representative of K. For $s \in \mathscr{S}$ we set

$$K_s = K \cap sKs^{-1} \tag{2.6}$$

and observe that given $g \in KsK$ we have $|\{(k_1, k_2) \in K^2 : k_1sk_2 = g\}| = |K_s|$. Indeed, suppose that $k_1sk_2 = g = h_1sh_2$, where $k_1, k_2, h_1, h_2 \in K$. Then we have ${h_1}^{-1}k_1 = sh_2{k_2}^{-1}s^{-1}$ which gives, in particular, ${h_1}^{-1}k_1 \in K_s$. Thus there are $|K_s| = |k_1K_s|$ different choices for h_1, and since $h_2 = s^{-1}{h_1}^{-1}g$ is determined by h_1, the observation follows.

As a consequence, given an Abelian group A (e.g. $\mathbb{C}$, a vector space, etc.), for any map $\Phi \colon G \to A$ and $s \in \mathscr{S}$ we have

$$\sum_{g \in KsK} \Phi(g) = \frac{1}{|K_s|} \sum_{k_1,k_2 \in K} \Phi(k_1sk_2). \tag{2.7}$$

For $s \in \mathscr{S}$ we denote by (θ^s, V_s) the K_s-representation defined by setting $V_s = V$ and

$$\theta^s(x) = \theta(s^{-1}xs) \tag{2.8}$$

for all $x \in K_s$.

For $T \in \mathrm{Hom}_{K_s}(\mathrm{Res}^K_{K_s}\theta, \theta^s)$ define $\mathscr{L}_T : G \to \mathrm{End}(V)$ by setting

$$\mathscr{L}_T(g) = \begin{cases} \theta(k_2^{-1})T\theta(k_1^{-1}) & \text{if } g = k_1sk_2 \text{ for some } k_1, k_2 \in K \\ 0 & \text{if } g \notin KsK. \end{cases} \tag{2.9}$$

Let us show that (2.9) is well defined and that $\mathscr{L}_T \in \widetilde{\mathscr{H}}(G, K, \theta)$. Let $k_1, k_2, h_1, h_2 \in K$.

Suppose again that $k_1sk_2 = g = h_1sh_2$, so that, as before, we can find $k_s \in K_s$ such that $k_1 = h_1k_s$ and therefore

$$k_1^{-1} = k_s{}^{-1}h_1{}^{-1} \text{ and } k_2^{-1} = h_2{}^{-1}s^{-1}k_ss.$$

We then have

$$\begin{aligned} \theta(k_2^{-1})T\theta(k_1^{-1}) &= \theta(h_2{}^{-1}s^{-1}k_ss)T\theta(k_s{}^{-1}h_1{}^{-1}) \\ &= \theta(h_2{}^{-1})\theta(s^{-1}k_ss)T\theta(k_s{}^{-1})\theta(h_1{}^{-1}) \\ &= \theta(h_2{}^{-1})\theta^s(k_s)T\theta(k_s{}^{-1})\theta(h_1{}^{-1}) \\ (\text{since } T \in \mathrm{Hom}_{K_s}(\mathrm{Res}^K_{K_s}\theta, \theta^s)) \quad &= \theta(h_2{}^{-1})T\theta(k_s)\theta(k_s{}^{-1})\theta(h_1{}^{-1}) \\ &= \theta(h_2{}^{-1})T\theta(h_1{}^{-1}). \end{aligned}$$

It follows that (2.9) is well defined.

Suppose now that $g = k_1sk_2$ so that $h_1gh_2 = h_1k_1sk_2h_2$. Then, by (2.9), we have

$$\begin{aligned} \mathscr{L}_T(h_1gh_2) &= \mathscr{L}_T(h_1k_1sk_2h_2) \\ &= \theta((k_2h_2)^{-1})T\theta((h_1k_1)^{-1}) \\ &= \theta(h_2{}^{-1})\left(\theta(k_2^{-1})T\theta(k_1^{-1})\right)\theta(h_1{}^{-1}) \\ &= \theta(h_2^{-1})\mathscr{L}_T(g)\theta(h_1^{-1}). \end{aligned}$$

This shows that $\mathscr{L}_T \in \widetilde{\mathscr{H}}(G, K, \theta)$.

We set

$$\mathscr{S}_0 = \{s \in \mathscr{S} : \mathrm{Hom}_{K_s}(\mathrm{Res}^K_{K_s}\theta, \theta^s) \text{ is nontrivial}\}. \tag{2.10}$$

Lemma 2.1

(1) *If* $F \in \widetilde{\mathscr{H}}(G, K, \theta)$ *then*

$$F(s) \in \mathrm{Hom}_{K_s}(\mathrm{Res}^K_{K_s}\theta, \theta^s) \textit{ for all } s \in \mathscr{S}. \tag{2.11}$$

(2) *If* $F \in \widetilde{\mathscr{H}}(G, K, \theta)$ *then*

$$F = \sum_{s \in \mathscr{S}_0} \mathscr{L}_{F(s)} \tag{2.12}$$

and the nontrivial elements in this sum are linearly independent.

(3) *If* $F_1, F_2 \in \widetilde{\mathscr{H}}(G, K, \theta)$ *then*

$$\langle F_1, F_2\rangle_{\widetilde{\mathscr{H}}(G,K,\theta)} = |K|^2 \sum_{s \in \mathscr{S}_0} \frac{1}{|K_s|}\langle F_1(s), F_2(s)\rangle_{\mathrm{End}(V)}.$$

Proof

(1) Let $s \in \mathscr{S}$. For all $x \in K_s$, by (2.1), we have

$$\begin{aligned} F(s)\theta(x) &= F(x^{-1}s) \\ &= F(s \cdot s^{-1}x^{-1}s) \\ &= \theta(s^{-1}xs)F(s) \\ &= \theta^s(x)F(s), \end{aligned}$$

that is, $F(s) \in \mathrm{Hom}_{K_s}(\mathrm{Res}^K_{K_s}\theta, \theta^s)$. In particular, $F(s) = 0$ if $s \notin \mathscr{S}_0$.

(2) From (2.1) we deduce $F(k_1sk_2) = \theta(k_2^{-1})F(s)\theta(k_1^{-1}) = \mathscr{L}_{F(s)}(k_1sk_2)$ for all $s \in S$ and $k_1, k_2 \in K$. As a consequence, F is determined by its values on $\mathscr{S}_0$. Moreover, for distinct $s, s' \in \mathscr{S}_0$ the maps $\mathscr{L}_{F(s)}$ and $\mathscr{L}_{F(s')}$ have disjoint supports (namely the double cosets KsK and $Ks'K$, respectively). From (2.5) we then deduce (2.12) and that the nontrivial elements in the sum are linearly independent.

(3) We have

$$\begin{aligned}
\langle F_1, F_2\rangle_{\widetilde{\mathscr{H}}(G,K,\theta)} &= \frac{1}{\dim V}\sum_{g\in G}\mathrm{tr}[F_2(g)^*F_1(g)]\\
\text{(by (2.5) and (2.7))} &= \frac{1}{\dim V}\sum_{s\in\mathscr{S}_0}\frac{1}{|K_s|}\sum_{k_1,k_2\in K}\mathrm{tr}[F_2(k_1sk_2)^*F_1(k_1sk_2)]\\
&= \frac{1}{\dim V}\sum_{s\in\mathscr{S}_0}\frac{1}{|K_s|}\sum_{k_1,k_2\in K}\mathrm{tr}[\theta(k_1)F_2(s)^*\theta(k_2)\theta(k_2^{-1})F_1(s)\theta(k_1^{-1})]\\
&= |K|^2\sum_{s\in\mathscr{S}_0}\frac{1}{|K_s|\cdot\dim V}\mathrm{tr}[F_2(s)^*F_1(s)]\\
&= |K|^2\sum_{s\in\mathscr{S}_0}\frac{1}{|K_s|}\langle F_1(s), F_2(s)\rangle_{\mathrm{End}(V)}.
\end{aligned}$$

□

Proposition 2.1 *For each* $s \in \mathscr{S}_0$, *select an orthonormal basis* $\{T_{s,1}, T_{s,2}, \ldots, T_{s,m_s}\}$ *in* $\mathrm{Hom}_{K_s}(\mathrm{Res}^K_{K_s}\theta, \theta^s)$. *Then the set* $\{\mathscr{L}_{T_{s,i}} : s \in \mathscr{S}_0, 1 \le i \le m_s\}$ *is an orthogonal basis of* $\widetilde{\mathscr{H}}(G, K, \theta)$ *and, for* $s, t \in \mathscr{S}_0$, $1 \le i \le m_s$, $1 \le j \le m_t$, *we have:*

$$\langle \mathscr{L}_{T_{s,i}}, \mathscr{L}_{T_{t,j}}\rangle_{\widetilde{\mathscr{H}}(G,K,\theta)} = \delta_{s,t}\delta_{i,j}\frac{|K|^2}{|K_s|}.$$

Proof From Lemma 2.1.(2) it follows that $\{\mathscr{L}_{T_{s,i}} : s \in \mathscr{S}_0, 1 \le i \le m_s\}$ is a basis of $\widetilde{\mathscr{H}}(G, K, \theta)$. The orthogonality relations follow easily from Lemma 2.1.(3). □

We now define a map

$$\xi\colon \widetilde{\mathscr{H}}(G, K, \theta) \to \mathrm{End}(\mathrm{Ind}_K^G V), \tag{2.13}$$

by setting

$$[\xi(F)f](g) = \sum_{h\in G} F(h^{-1}g)f(h) \tag{2.14}$$

for all $F \in \widetilde{\mathscr{H}}(G, K, \theta)$, $f \in \mathrm{Ind}_K^G V$ and $g \in G$. Note that $F(h^{-1}g)f(h)$ indicates the action of the operator $F(h^{-1}g)$ on the vector $f(h)$.

Lemma 2.2 *If* $F \in \widetilde{\mathscr{H}}(G, K, \theta)$ *then*

$$\mathrm{tr}[\xi(F)] = |G| \cdot \mathrm{tr}[F(1_G)].$$

Proof Suppose that f_v is as in (1.22) and $t \in \mathcal{T}$ (see (1.21)). We then have:

$$\begin{aligned}
[\xi(F)\lambda(t)f_v](g) &= \sum_{h\in G} F(h^{-1}g)f_v(t^{-1}h) \\
(\text{setting } t^{-1}h = k) \quad &= \sum_{k\in K} F(k^{-1}t^{-1}g)\theta(k^{-1})v \qquad (2.15)\\
(\text{by } (2.1)) \quad &= |K|F(t^{-1}g)v.
\end{aligned}$$

Then, computing the trace of $\xi(F)$ by means of the orthonormal basis (1.25) we get:

$$\begin{aligned}
\mathrm{tr}[\xi(F)] &= \sum_{t\in\mathcal{T}}\sum_{j=1}^{d_\theta} \langle \xi(F)\lambda(t)f_{v_j}, \lambda(t)f_{v_j}\rangle_{\mathrm{Ind}_K^G V} \\
(\text{by } (1.24)) \quad &= \sum_{t\in\mathcal{T}}\sum_{j=1}^{d_\theta} \frac{1}{|K|}\sum_{g\in G} \langle [\xi(F)\lambda(t)f_{v_j}](g), [\lambda(t)f_{v_j}](g)\rangle_V \\
(\text{by } (2.15)) \quad &= \sum_{t\in\mathcal{T}}\sum_{j=1}^{d_\theta}\sum_{g\in G} \langle F(t^{-1}g)v_j, f_{v_j}(t^{-1}g)\rangle_V \\
(\text{setting } t^{-1}g = k) \quad &= \sum_{t\in\mathcal{T}}\sum_{j=1}^{d_\theta}\sum_{k\in K} \langle F(k)v_j, \theta(k^{-1})v_j\rangle_V \\
(\text{by } (2.1)) \quad &= \sum_{t\in\mathcal{T}}\sum_{j=1}^{d_\theta} |K|\langle F(1_G)v_j, v_j\rangle_V \\
&= |G|\mathrm{tr}[F(1_G)].
\end{aligned}$$

□

We are now in a position to state and prove the main result of this section.

Theorem 2.1 $\xi(F) \in \mathrm{End}_G(\mathrm{Ind}_K^G V)$ *for all* $F \in \widetilde{\mathcal{H}}(G,K,\theta)$ *and* ξ *is a* $*$-*isomorphism between* $\widetilde{\mathcal{H}}(G,K,\theta)$ *and* $\mathrm{End}_G(\mathrm{Ind}_K^G V)$. *Moreover,*

$$\langle \xi(F_1), \xi(F_2)\rangle_{\mathrm{End}_G(\mathrm{Ind}_K^G V)} = |K|\langle F_1, F_2\rangle_{\widetilde{\mathcal{H}}(G,K,\theta)}, \qquad (2.16)$$

for all $F_1, F_2 \in \widetilde{\mathcal{H}}(G,K,\theta)$, *that is, the normalized map* $F \mapsto \frac{1}{\sqrt{|K|}}\xi(F)$ *is an isometry.*

Proof Let $F \in \mathscr{H}(G, K, \theta)$, $f \in \mathrm{Ind}_K^G V$ and $g, h \in G$. Then, if $\lambda = \mathrm{Ind}_K^G \theta$ as before, we have

$$
\begin{aligned}
[\lambda(h)\xi(F)f](g) &= [\xi(F)f](h^{-1}g)\\
\text{(by (2.14))} \quad &= \sum_{r\in G} F(r^{-1}h^{-1}g)f(r)\\
\text{(setting } q = hr) \quad &= \sum_{q\in G} F(q^{-1}g)f(h^{-1}q)\\
&= \sum_{q\in G} F(q^{-1}g)[\lambda(h)f](q)\\
&= [\xi(F)\lambda(h)f](g),
\end{aligned}
$$

that is, $\lambda(h)\xi(F) = \xi(F)\lambda(h)$. This shows that $\xi(F) \in \mathrm{End}_G(\mathrm{Ind}_K^G V)$. Moreover it is also easy to check that

$$
\xi(F_1 * F_2) = \xi(F_1)\xi(F_2). \tag{2.17}
$$

and

$$
\xi(F)^* = \xi(F^*). \tag{2.18}
$$

Just note that $\xi(F)^*$ is the adjoint of an operator in $\mathrm{End}(\mathrm{Ind}_K^G V)$ with respect to the scalar product in (1.24). By (2.17) and (2.18) we have $\xi(F_2)^*\xi(F_1) = \xi(F_2^* * F_1)$ and therefore

$$
\begin{aligned}
\langle \xi(F_1), \xi(F_2)\rangle_{\mathrm{End}_G(\mathrm{Ind}_K^G V)} &= \frac{1}{\dim V \cdot |G/K|}\mathrm{tr}[\xi(F_2^* * F_1)]\\
\text{(by Lemma 2.2)} \quad &= \frac{|K|}{\dim V}\mathrm{tr}[(F_2^* * F_1)(1_G)]\\
&= \frac{|K|}{\dim V}\sum_{g\in G}\mathrm{tr}[F_2^*(g^{-1})F_1(g)]\\
&= \frac{|K|}{\dim V}\sum_{g\in G}\mathrm{tr}[F_2(g)^*F_1(g)]\\
\text{(by (2.3))} \quad &= |K|\langle F_1, F_2\rangle_{\mathscr{H}(G,K,\theta)},
\end{aligned}
$$

and (2.16) follows. In particular, ξ is injective. It only remains to prove that ξ is surjective.

Let $T \in \mathrm{End}_G(\mathrm{Ind}_K^G V)$ and define $\Xi(T): G \to \mathrm{End}(V)$ by setting, for all $v \in V$ and $g \in G$,

$$\Xi(T)(g)v = \frac{1}{|K|}[Tf_v](g), \tag{2.19}$$

where f_v is as in (1.22). For $k_1, k_2 \in K$ we then have

$$\begin{aligned}
\Xi(T)(k_1gk_2)v &= \frac{1}{|K|}[Tf_v](k_1gk_2) \\
(\text{since } Tf_v \in \mathrm{Ind}_K^G V) \quad &= \theta(k_2^{-1})\frac{1}{|K|}[Tf_v](k_1g) \\
&= \theta(k_2^{-1})\left\{\frac{1}{|K|}\lambda(k_1^{-1})[Tf_v](g)\right\} \\
(\text{since } T \in \mathrm{End}_G(\mathrm{Ind}_K^G V)) \quad &= \theta(k_2^{-1})\left\{\frac{1}{|K|}[T\lambda(k_1^{-1})f_v](g)\right\} \\
(\text{since } \lambda(k_1^{-1})f_v = f_{\theta(k_1^{-1})v}) \quad &= \theta(k_2^{-1})\left\{\frac{1}{|K|}[Tf_{\theta(k_1^{-1})v}](g)\right\} \\
&= \theta(k_2^{-1})[\Xi(T)(g)]\theta(k_1^{-1})v.
\end{aligned}$$

This shows that $\Xi(T) \in \widetilde{\mathscr{H}}(G, K, \theta)$.

Let us show that the map $\Xi: \mathrm{End}_G(\mathrm{Ind}_K^G V) \to \widetilde{\mathscr{H}}(G, K, \theta)$ is a right-inverse of ξ. First observe that for all $f \in \mathrm{Ind}_K^G V$

$$\frac{1}{|K|}\sum_{h \in G}[\lambda(h)f_{f(h)}] = f, \tag{2.20}$$

which is an equivalent form of (1.23). Indeed, for all $g \in G$ we have

$$\begin{aligned}
\frac{1}{|K|}\sum_{h \in G}[\lambda(h)f_{f(h)}](g) &= \frac{1}{|K|}\sum_{h \in G}[f_{f(h)}](h^{-1}g) \\
&= \frac{1}{|K|}\sum_{h \in G}\left(\begin{cases}\theta(g^{-1}h)f(h) & \text{if } h^{-1}g \in K \\ 0 & \text{otherwise}\end{cases}\right) \\
&= \frac{1}{|K|}\sum_{k \in K}\left(\begin{cases}\theta(k)f(gk) = f(g) & \text{if } h^{-1}g = k^{-1} \in K \\ 0 & \text{otherwise}\end{cases}\right) \\
(\text{by (1.19)}) \quad &= f(g).
\end{aligned}$$

Let then $T \in \mathrm{End}_G(\mathrm{Ind}_K^G V)$, $f \in \mathrm{Ind}_K^G$ and $g \in G$. We have

$$\begin{aligned}
[(\xi \circ \Xi(T)f](g) &= \sum_{h \in G} [\Xi(T)(h^{-1}g)]f(h) \\
\text{(by (2.19))} &= \frac{1}{|K|} \sum_{h \in G} \left[T f_{f(h)}\right](h^{-1}g) \\
&= \frac{1}{|K|} \sum_{h \in G} \left[\lambda(h) T f_{f(h)}\right](g) \\
\text{(since } T \in \mathrm{End}_G(\mathrm{Ind}_K^G V)) &= \frac{1}{|K|} \sum_{h \in G} \left[T\lambda(h) f_{f(h)}\right](g) \\
&= \left[T\left(\frac{1}{|K|}\sum_{h \in G} \lambda(h) f_{f(h)}\right)\right](g) \\
\text{(by (2.20))} &= [Tf](g).
\end{aligned}$$

This shows that $\xi(\Xi(T)) = T$, and surjectivity of ξ follows. □

Recalling Proposition 2.1, we immediately get the following:

Corollary 2.1 (Mackey's Intertwining Number Theorem) *Let (θ, V) be a K-representation. Then*

$$\dim \mathrm{End}_G(\mathrm{Ind}_K^G V) = \sum_{s \in \mathscr{S}_0} \dim \mathrm{Hom}_{K_s}(\mathrm{Res}_{K_s}^K \theta, \theta^s).$$

Remark 2.1 Now we prove that $\Xi(T)$ given by (2.19) is the unique ξ^{-1}-image of $T \in \mathrm{End}(\mathrm{Ind}_K^G V)$ (note that this yields a second proof of injectivity of ξ). Since the functions $\lambda(h) f_v$, $h \in G$ and $v \in V$, span $\mathrm{Ind}_K^G V$, for $F_1 \in \widetilde{\mathscr{H}}(G, K, \theta)$ we have that $\xi(F_1) = T$ if and only if

$$\xi(F_1)\lambda(h) f_v = T\lambda(h) f_v \quad \text{for all } h \in G \text{ and } v \in V. \tag{2.21}$$

Let $\Xi(T)$ be given by (2.19). If (2.21) holds, then

$$\begin{aligned}
|K|\Xi(T)(g)v &= [Tf_v](g) \\
&= [\lambda(g^{-1})Tf_v](1_G) \\
\text{(since } T \in \mathrm{End}_G(\mathrm{Ind}_K^G V)) &= [T\lambda(g^{-1})f_v](1_G) \\
\text{(by (2.21))} &= [\xi(F_1)\lambda(g^{-1})f_v](1_G)
\end{aligned}$$

$$
\begin{aligned}
\text{(by (2.14))} &= \sum_{h\in G} F_1(h^{-1}) f_v(gh) \\
\text{(by (1.22)) with } gh = k) &= \sum_{k\in K} F_1(k^{-1}g)\theta(k^{-1})v \\
\text{(by (2.1))} &= |K| F_1(g)v.
\end{aligned}
$$

This shows that $F_1 = \Xi(T)$, that is, (2.19) defines the unique element in $\xi^{-1}(T)$.

We end this section by giving an explicit formula for the ξ-image of the map $\mathscr{L}_T \in \widetilde{\mathscr{H}}(G, K, \theta)$, where $T \in \mathrm{Hom}_{K_s}(\mathrm{Res}^K_{K_s}\theta, \theta^s)$ and $s \in \mathscr{S}$, defined in (2.9), and therefore, for the elements in the orthogonal basis of $\widetilde{\mathscr{H}}(G, K, \theta)$ (cf. Proposition 2.1).

Fix $s \in \mathscr{S}$ and choose a transversal $\mathscr{R}_s$ for the right-cosets of $K_{s^{-1}} \equiv K \cap s^{-1}Ks = s^{-1}K_s s$ in K so that $K = \bigsqcup_{r\in\mathscr{R}_s} K_{s^{-1}}r$. Then, for all $f \in \mathrm{Ind}_K^G V$ and $g \in G$, we have

$$
\begin{aligned}
[\xi(\mathscr{L}_T)f](g) &= \sum_{h\in G} \mathscr{L}_T(h^{-1}g) f(h) \\
(z = h^{-1}g) &= \sum_{z\in G} \mathscr{L}_T(z) f(gz^{-1}) \\
\text{(by (2.7) and (2.9))} &= \frac{1}{|K_s|} \sum_{k,k_1\in K} \theta(k^{-1})T\theta(k_1^{-1}) f(gk^{-1}s^{-1}k_1^{-1}) \\
\text{(by (1.19))} &= \frac{|K|}{|K_s|} \sum_{k\in K} \theta(k^{-1})Tf(gk^{-1}s^{-1}) \\
(k = xr) &= \frac{|K|}{|K_s|} \sum_{x\in K_{s^{-1}}} \sum_{r\in\mathscr{R}_s} \theta(r^{-1})\theta(x^{-1})Tf(gr^{-1}x^{-1}s^{-1}) \\
\text{(by (2.8) and } sxs^{-1} \in K_s) &= \frac{|K|}{|K_s|} \sum_{x\in K_{s^{-1}}} \sum_{r\in\mathscr{R}_s} \theta(r^{-1})\theta^s(sx^{-1}s^{-1}) \\
&\quad \cdot Tf(gr^{-1}x^{-1}s^{-1}) \\
(T \in \mathrm{Hom}_{K_s}(\mathrm{Res}^K_{K_s}\theta, \theta^s)) &= \frac{|K|}{|K_s|} \sum_{x\in K_{s^{-1}}} \sum_{r\in\mathscr{R}_s} \theta(r^{-1})T\theta(sx^{-1}s^{-1}) \\
&\quad \cdot f(gr^{-1}s^{-1}(sx^{-1}s^{-1})) \\
\text{(by (1.19))} &= |K| \sum_{r\in\mathscr{R}_s} \theta(r^{-1})Tf(gr^{-1}s^{-1}).
\end{aligned}
$$

2.2 The Hecke Algebra Revisited

Let (θ, V) be a K-representation as in the previous section, but we now assume that θ is *irreducible*. We choose $v \in V$ with $\|v\| = 1$ and an orthonormal basis $\{v_j : j = 1, 2, \ldots, d_\theta\}$ in V with $v_1 = v$. We begin with a technical but quite useful lemma.

Lemma 2.3 *For all $w \in V$ and $1 \le i, j \le d_\theta$, we have:*

$$\frac{d_\theta}{|K|} \sum_{k \in K} \langle \theta(k)v_i, v_j \rangle \theta(k^{-1})w = \langle w, v_j \rangle v_i. \tag{2.22}$$

Proof Let $w = \sum_{r=1}^{d_\theta} \alpha_r v_r$. Then, for all $1 \le s \le d_\theta$, we have

$$\begin{aligned}\left\langle \frac{d_\theta}{|K|} \sum_{k \in K} \langle \theta(k)v_i, v_j \rangle \theta(k^{-1})w, v_s \right\rangle &= \sum_{r=1}^{d_\theta} \alpha_r \frac{d_\theta}{|K|} \sum_{k \in K} \langle \theta(k)v_i, v_j \rangle \langle v_r, \theta(k)v_s \rangle \\ \text{(by (1.14))} &= \sum_{r=1}^{d_\theta} \alpha_r \delta_{i,s} \delta_{j,r} \\ &= \langle w, v_j \rangle \delta_{i,s}\end{aligned}$$

so that the left hand side of (2.22) is equal to

$$\sum_{s=1}^{d_\theta} \langle w, v_j \rangle \delta_{i,s} v_s = \langle w, v_j \rangle v_i. \qquad \square$$

In the sequel, we shall use several times the following particular case (obtained from (2.22) after takig $w = v_j$):

$$v_i = \frac{d_\theta}{|K|} \sum_{k \in K} \langle \theta(k)v_i, v_j \rangle_V \theta(k^{-1})v_j. \tag{2.23}$$

Define $\psi \in L(K) = \{f \colon K \to \mathbb{C}\}$ by setting

$$\psi(k) = \frac{d_\theta}{|K|} \langle v, \theta(k)v \rangle_V \tag{2.24}$$

for all $k \in K$. From now on, we identify $L(K)$ as the subalgebra of $L(G)$ consisting of all functions supported on K. Thus, we regard ψ also as an element in $L(G)$. We then define the convolution operator $P \colon L(G) \to L(G)$ by setting

$$Pf = f * \psi$$

for all $f \in L(G)$ (in fact, $P = T_\psi$, cf. (1.7)). We also define the operator

$$T_v \colon \mathrm{Ind}_K^G V \to L(G)$$

by setting

$$[T_v f](g) = \sqrt{d_\theta/|K|}\langle f(g), v\rangle_V \tag{2.25}$$

for all $f \in \mathrm{Ind}_K^G V$ and $g \in G$, and denote its range by

$$\mathscr{I}(G, K, \psi) = T_v\left(\mathrm{Ind}_K^G V\right) \subseteq L(G).$$

Finally, we define a map

$$S_v \colon \widetilde{\mathscr{H}}(G, K, \theta) \longrightarrow L(G)$$

by setting

$$[S_v F](g) = d_\theta \langle F(g)v, v\rangle_V$$

for all $F \in \widetilde{\mathscr{H}}(G, K, \theta)$ and $g \in G$. The first and the second statement in the following theorem are taken from [61]. We reproduce here the proofs for the sake of completeness.

Theorem 2.2

(1) *The operator T_v belongs to $\mathrm{Hom}_G(\mathrm{Ind}_K^G V, L(G))$ and it is an isometry; in particular, $\mathscr{I}(G, K, \psi)$ is a λ_G-invariant subspace of $L(G)$, which is G-isomorphic to $\mathrm{Ind}_K^G V$.*

(2) *The function ψ satisfies the identities*

$$\psi * \psi = \psi \text{ and } \psi^* = \psi; \tag{2.26}$$

moreover, P is the orthogonal projection of $L(G)$ onto $\mathscr{I}(G, K, \psi)$. In other words,

$$\mathscr{I}(G, K, \psi) = \{f * \psi : f \in L(G)\} \equiv \{f \in L(G) : f * \psi = f\}. \tag{2.27}$$

(3) *Define*

$$\mathscr{H}(G, K, \psi) = \{\psi * f * \psi : f \in L(G)\} \equiv \{f \in L(G) : f = \psi * f * \psi\}.$$

Then $\mathscr{H}(G, K, \psi)$ is an involutive subalgebra of $\mathscr{I}(G, K, \psi) \subseteq L(G)$ and S_v yields a $*$-antiisomorphism from $\widetilde{\mathscr{H}}(G, K, \theta)$ onto $\mathscr{H}(G, K, \psi)$. Every $f \in \mathscr{H}(G, K, \psi)$ is supported in $\bigsqcup_{s \in \mathscr{S}_0} KsK$ (cf. (2.10)). Moreover, $\frac{1}{\sqrt{d_\theta}} S_v$ is an *isometry and we have*

$$T_v\,[\xi(F)f] = (T_v f) * (S_v F) \tag{2.28}$$

for all $f \in \mathrm{Ind}_K^G V$ *and* $F \in \widetilde{\mathscr{H}}(G, K, \theta)$. *The inverse* S_v^{-1} *of* S_v *is given by*

$$\langle [S_v^{-1} f](g) v_i, v_j\rangle = \frac{d_\theta}{|K|^2} \sum_{k_1,k_2\in K} f(k_1^{-1} g k_2)\overline{\langle \theta(k_1)v_1, v_i\rangle \langle \theta(k_2)v_1, v_j\rangle} \tag{2.29}$$

for all $f \in \mathscr{H}(G, K, \psi)$, $g \in G$, *and* $i, j = 1, 2, \ldots, d_\theta$.

(4) *The map*

$$\begin{array}{ccc} \mathscr{H}(G, K, \psi) & \longrightarrow & \mathrm{End}_G(\mathscr{I}(G, K, \psi)) \\ f & \longmapsto & T_f|_{\mathscr{I}(G,K,\psi)} \end{array} \tag{2.30}$$

is a $*$*-antiisomorphism of algebras and* $\ker T_f$ *contains* $\{\mathscr{I}(G, K, \psi)\}^\perp$ *for all* $f \in \mathscr{H}(G, K, \psi)$. *Moreover,*

$$T_f|_{\mathscr{I}(G,K,\psi)} = T_v \circ \xi\left(S_v^{-1} f\right) \circ T_v{}^{-1} \tag{2.31}$$

and

$$\langle T_{f_1}, T_{f_2}\rangle_{\mathrm{End}_G(\mathscr{I}(G,K,\psi))} = \frac{|G|}{\dim \mathscr{I}(G, K, \psi)} \langle f_1, f_2\rangle_{L(G)} \tag{2.32}$$

for all $f, f_1, f_2 \in \mathscr{H}(G, K, \psi)$.

Before starting the proof, for the convenience of the reader, in diagrams

$$\mathrm{Ind}_K^G V \xrightarrow{T_v} \mathscr{I}(G, K, \psi)$$

and

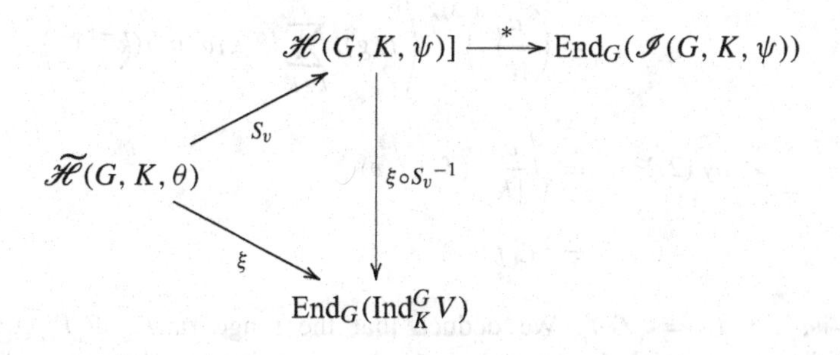

where $\xrightarrow{*}$ denotes the map defined in (2.30) and (2.31), we summarize the various maps involved in the theorem.

Proof

(1) It is immediate to check that

$$T_v\lambda(g)f = \lambda_G(g)T_v f$$

for all $g \in G$ and $f \in \mathrm{Ind}_K^G V$, showing that $T_v \in \mathrm{Hom}_G(\mathrm{Ind}_K^G V, L(G))$. We now show that T_v is an isometry by using the basis in (1.25) (recall that $\mathscr{T}$ is a transversal for the left-cosets of K in G). For $t_1, t_2 \in \mathscr{T}$ and $i, j = 1, 2, \ldots, d_\theta$, we have

$$\begin{aligned}\langle T_v\lambda(t_1)f_{v_i}, T_v\lambda(t_2)f_{v_j}\rangle_{L(G)} &= \frac{d_\theta}{|K|}\sum_{g\in G}\langle f_{v_i}(t_1^{-1}g), v\rangle\overline{\langle f_{v_j}(t_2^{-1}g), v\rangle}\\ \text{(by (1.22))} \quad &= \frac{d_\theta}{|K|}\delta_{t_1,t_2}\sum_{k\in K}\langle\theta(k^{-1})v_i, v_1\rangle\overline{\langle\theta(k^{-1})v_j, v_1\rangle}\\ \text{(by (1.14))} \quad &= \delta_{t_1,t_2}\delta_{i,j}\\ &= \langle\lambda(t_1)f_{v_i}, \lambda(t_2)f_{v_j}\rangle_{L(G)}.\end{aligned}$$

(2) The first identity in (2.26) follows from the convolution properties of matrix coefficients ((1.15) with $\sigma = \rho = \theta$ and $i = j = h = k = 1$) and ensures that P is an idempotent; the second identity in (2.26) follows from the fact that ψ is (the conjugate of) a diagonal matrix coefficient, and (1.9) ensures that P is self-adjoint. As a consequence, $P = P^2 = P^*$ is an orthogonal projection. Moreover, for all $f \in \mathrm{Ind}_K^G V$ and $g \in G$ we have

$$\begin{aligned}[(T_v f) * \psi](g) &= \left(\frac{d_\theta}{|K|}\right)^{3/2}\sum_{k\in K}\langle f(gk^{-1}), v\rangle\langle v, \theta(k)v\rangle\\ \text{(by (1.19))} \quad &= \left(\frac{d_\theta}{|K|}\right)^{3/2}\sum_{k\in K}\langle\theta(k)f(g), v\rangle\langle v, \theta(k)v\rangle\\ &= \left(\frac{d_\theta}{|K|}\right)^{3/2}\left\langle f(g), \sum_{k\in K}\langle\theta(k)v, v\rangle\theta(k^{-1})v\right\rangle\\ \text{(by (2.23))} \quad &= \sqrt{\frac{d_\theta}{|K|}}\langle f(g), v\rangle\\ &= (T_v f)(g),\end{aligned}$$

that is, $PT_v f = T_v f$. We deduce that the range $\mathrm{ran}P$ of P contains $T_v(\mathrm{Ind}_K^G V) = \mathscr{I}(G, K, \psi)$. Let now show that, in fact, the range of P is contained in (and therefore equals) $\mathscr{I}(G, K, \psi)$. Indeed, for all $\phi \in L(G)$

and $g \in G$ we have

$$\begin{aligned}[P\phi](g) &= \sum_{k\in K} \phi(gk)\psi(k^{-1}) \\ &= \frac{d_\theta}{|K|}\left\langle \sum_{k\in K} \phi(gk)\theta(k)v, v\right\rangle \\ &= [T_v f](g)\end{aligned}$$

where $f: G \to V$ is defined by $f(g) = \sqrt{\frac{d_\theta}{|K|}}\sum_{k\in K}\phi(gk)\theta(k)v$. Since f belongs to $\mathrm{Ind}_K^G V$, as one immediately checks, we conclude that $\mathrm{ran}P = T_v(\mathrm{Ind}_K^G V)$.

(3) It is easy to check that $\mathscr{H}(G, K, \psi) \subseteq \mathscr{I}(G, K, \psi)$ and that $\mathscr{H}(G, K, \psi)$ is closed under the convolution product and the involution $f \mapsto f^*$. As a consequence, $\mathscr{H}(G, K, \psi)$ is a $*$-subalgebra of $L(G)$ contained in $\mathscr{I}(G, K, \psi)$.

Now we prove the statements for S_v. Let $F, F_1, F_2 \in \widetilde{\mathscr{H}}(G, K, \theta)$, $f \in \mathrm{Ind}_K^G V$ and $g \in G$. We then have

$$\begin{aligned}[\psi * (S_v F) * \psi](g) &= \frac{d_\theta^3}{|K|^2}\sum_{k_1,k_2\in K}\langle v_1, \theta(k_1)v_1\rangle\langle F(k_1^{-1}gk_2^{-1})v_1, v_1\rangle \\ &\quad\cdot\langle v_1, \theta(k_2)v_1\rangle \\ \text{(by (2.1))}\quad &= \frac{d_\theta^3}{|K|^2}\Bigg\langle F(g)\sum_{k_1\in K}\langle\theta(k_1^{-1})v_1, v_1\rangle\theta(k_1)v_1, \\ &\quad \sum_{k_2\in K}\langle\theta(k_2)v_1, v_1\rangle\theta(k_2^{-1})v_1\Bigg\rangle \\ \text{(by (2.23))}\quad &= (S_v F)(g)\end{aligned}$$

and therefore $S_v\left[\widetilde{\mathscr{H}}(G, K, \theta)\right] \subseteq \mathscr{H}(G, K, \psi)$. Moreover

$$\begin{aligned}\langle S_v(F_1), S_v(F_2)\rangle_{L(G)} &= d_\theta^2\sum_{g\in G}\langle F_1(g)v, v\rangle\overline{\langle F_2(g)v, v\rangle} \\ \text{(by (2.23))} &= \frac{d_\theta^2}{|K|^2}\sum_{g\in G}\sum_{i,j=1}^{d_\theta}\sum_{k_1,k_2\in K}\langle F_1(g)\theta(k_1^{-1})v_i, \theta(k_2^{-1})v_j\rangle\langle\theta(k_1)v_1, v_i\rangle \\ &\quad\cdot\overline{\langle\theta(k_2)v_1, v_j\rangle\langle F_2(g)v_1, v_1\rangle}\end{aligned}$$

$$(h = k_1 g k_2^{-1}) = \frac{d_\theta^2}{|K|^2} \sum_{h\in G} \sum_{i,j=1}^{d_\theta} \langle F_1(h)v_i, v_j\rangle$$

$$\overline{\cdot \left\langle F_2(h) \sum_{k_1\in K} \langle\theta(k_1^{-1})v_i, v_1\rangle\theta(k_1)v_1, \sum_{k_2\in K} \langle\theta(k_2^{-1})v_j, v_1\rangle\theta(k_2)v_1 \right\rangle}$$

$$(\text{by } (2.23)) = \sum_{h\in G} \sum_{i,j=1}^{d_\theta} \langle F_1(h)v_i, v_j\rangle\overline{\langle F_2(h)v_i, v_j\rangle}$$

$$= \sum_{i=1}^{d_\theta} \sum_{h\in G} \langle F_2(h)^* F_1(h)v_i, v_i\rangle$$

$$= d_\theta \langle F_1, F_2\rangle_{\widetilde{\mathscr{H}}(G,K,\theta)}$$

and this proves that $\frac{1}{\sqrt{d_\theta}} S_v$ is an isometry and, in particular, that S_v is injective.

Given $f \in L(G)$, we define $F\colon G \longrightarrow \text{End}(V)$ by setting

$$\langle F(g)v_i, v_j\rangle = \frac{d_\theta}{|K|^2} \sum_{k_1,k_2\in K} f(k_1^{-1} g k_2)\overline{\langle\theta(k_1)v_1, v_i\rangle}\langle\theta(k_2)v_1, v_j\rangle$$

for all $g \in G$ and $i, j = 1, 2, \ldots, d_\theta$. Then it is immediate to check that $F \in \widetilde{\mathscr{H}}(G, K, \theta)$. Moreover, if in addition $f \in \mathscr{H}(G, K, \psi)$, we then have, for all $g \in G$,

$$\begin{aligned} d_\theta\langle F(g)v, v\rangle &= \frac{d_\theta^2}{|K|^2} \sum_{k_1,k_2\in K} \langle v, \theta(k_1)v\rangle f(k_1^{-1} g k_2)\langle v, \theta(k_2^{-1})v\rangle \\ &= (\psi * f * \psi)(g) = f(g), \end{aligned}$$

that is, $S_v F = f$. This shows that S_v is also surjective (in (4) another proof of surjectivity is given). Incidentally, the above also provides the expression (2.29) for S_v^{-1}.

It is easy to check that S_v preserves the involutions: indeed,

$$(S_v F^*)(g) = d_\theta\langle F(g^{-1})^* v, v\rangle = d_\theta\overline{\langle F(g^{-1})v, v\rangle} = (S_v F)^*(g),$$

where the first equality follows from (2.4).

We need a little more effort to prove that S_v is an antiisomorphism:

$$
\begin{aligned}
[S_v(F_1 * F_2)](g) &= d_\theta \sum_{h\in G} \langle F_1(h^{-1}g)F_2(h)v, v\rangle \\
&= d_\theta \sum_{h\in G}\sum_{i=1}^{d_\theta} \langle F_2(h)v_1, v_i\rangle\langle F_1(h^{-1}g)v_i, v_1\rangle \\
\text{(by (2.23))} \quad &= \frac{d_\theta^2}{|K|} \sum_{h\in G}\sum_{i=1}^{d_\theta}\sum_{k\in K} \langle F_2(h)v_1, \theta(k^{-1})v_1\rangle\langle v_1, \theta(k)v_i\rangle \\
&\quad \cdot \langle F_1(h^{-1}g)v_i, v_1\rangle \\
(h = rk) \quad &= \frac{d_\theta^2}{|K|} \sum_{r\in G}\sum_{i=1}^{d_\theta} \langle F_2(r)v_1, v_1\rangle \Big\langle F_1(r^{-1}g) \\
&\quad \cdot \sum_{k\in K} \langle \theta(k^{-1})v_1, v_i\rangle \theta(k)v_i, v_1 \Big\rangle \\
\text{(by (2.23))} \quad &= [S_v(F_2) * S_v(F_1)]\,(g).
\end{aligned}
$$

This ends the proof that S_v is a $*$-antiisomorphism.

The fact that every $f \in \mathscr{H}(G, K, \psi)$ is supported in $\bigsqcup_{s\in\mathscr{S}_0} KsK$ follows immediately from the (anti-) isomorphsm between $\widetilde{\mathscr{H}}(G, K, \theta)$ and $\mathscr{H}(G, K, \psi)$ and Proposition 2.1.

We now prove (2.28):

$$
\begin{aligned}
[T_v\xi(F)f](g) &= \sqrt{\frac{d_\theta}{|K|}} \sum_{h\in G} \langle F(h^{-1}g)f(h), v\rangle \\
&= \sqrt{\frac{d_\theta}{|K|}} \sum_{i=1}^{d_\theta}\sum_{h\in G} \langle f(h), v_i\rangle\langle F(h^{-1}g)v_i, v_1\rangle \\
\text{(by (2.23))} \quad &= \left(\frac{d_\theta}{|K|}\right)^{3/2} \sum_{i=1}^{d_\theta}\sum_{h\in G}\sum_{k\in K} \langle f(h), \theta(k^{-1})v_1\rangle\langle v_1, \theta(k)v_i\rangle \\
&\quad \cdot \langle F(h^{-1}g)v_i, v_1\rangle \\
(h = rk) \quad &= \left(\frac{d_\theta}{|K|}\right)^{3/2} \sum_{i=1}^{d_\theta}\sum_{r\in G} \langle f(r), v\rangle \\
&\quad \cdot \Big\langle F(r^{-1}g) \sum_{k\in K} \langle \theta(k^{-1})v_1, v_i\rangle\theta(k)v_i, v_1 \Big\rangle \\
\text{(by (2.23))} \quad &= (T_v f) * (S_v F)(g).
\end{aligned}
$$

(4) Let $f \in \mathscr{H}(G, K, \psi)$ and $f_1 \in L(G)$. Then, if $f_1 = Pf_1$, that is $f_1 = f_1 * \psi$, we have $P(T_f f_1) = P(f_1 * f) = ((f_1 * \psi) * f) * \psi = f_1 * f = T_f f_1$. In other words $PT_f P = T_f P$, equivalently $PT_f|_{\mathscr{I}(G,K,\psi)} = T_f P|_{\mathscr{I}(G,K,\psi)}$. On the other hand, if $f_1 \in \ker P$ we have $T_f f_1 = f_1 * f = f_1 * (\psi * f * \psi) = (f_1 * \psi) * f * \psi = (Pf_1) * f * \psi = 0$, that is, $T_f \ker P \subseteq \ker P$.

This shows that the convolution operator T_f intertwines $\mathscr{I}(G, K, \psi)$ with itself and annihilates the orthogonal complement of $\mathscr{I}(G, K, \psi)$. This can be used to give a second proof of the surjectivity of S_v. Indeed, the fundamental relation (2.28) and the injectivity of S_v ensure that the antiisomorphism S_v identifies the commutant of $\mathscr{I}(G, K, \psi)$ with the convolution algebra $\mathscr{H}(G, K, \psi)$, so that $\dim \mathscr{H}(G, K, \psi) = \dim \mathrm{End}_G(\mathrm{Ind}_K^G V) \equiv \dim \widetilde{\mathscr{H}}(G, K, \theta)$. Moreover, keeping in mind (1.10), if $\bar{f} \in \mathrm{Ind}_K^G V$ we have

$$\begin{aligned}\left[T_v \circ \xi\left(S_v^{-1} f\right)\right](\bar{f}) &= T_v\left(\xi\left(S_v^{-1} f\right)\bar{f}\right)\\ \text{(by (2.28))} \quad &= T_v \bar{f} * f\\ &= \left[T_f \circ T_v\right](\bar{f})\end{aligned}$$

which gives (2.31).

Finally, (2.32) follows from (1.12). □

We end this section with a useful computational rule.

Lemma 2.4 *Let $f_1 \in \mathscr{H}(G, K, \psi)$ and $f_2 \in L(G)$. Then*

$$[f_1 * \psi * f_2 * \psi](1_G) = [f_1 * f_2](1_G). \tag{2.33}$$

Proof We have

$$\begin{aligned}[f_1 * \psi * f_2 * \psi](1_G) &= \sum_{h \in G}[f_1 * \psi * f_2](h)\psi(h^{-1})\\ &= \sum_{h \in G} \psi(h^{-1})[f_1 * \psi * f_2](h)\\ &= [\psi * f_1 * \psi * f_2](1_G)\\ &= [f_1 * f_2](1_G).\end{aligned}$$

□

Remark 2.2 In the terminology of [20, 22, 60], the function ψ is a *primitive idempotent* in $L(K)$ and $\mathscr{I}(G, K, \psi) \cap L(K) = \{f \in L(K) : f * \psi = f\}$ is the *minimal left ideal* in $L(K)$ generated by ψ (for, $\mathscr{I}(G, K, \psi)$ is generated by ψ as a left ideal in $L(G)$). Moreover, $\mathscr{H}(G, K, \psi)$ is the *Hecke algebra* associated with G, K and the primitive idempotent ψ.

Remark 2.3 For an element $\mathscr{L}_T$ of $\widetilde{\mathscr{H}}(G, K, \theta)$ as in (2.9) we have:

$$(S_v \mathscr{L}_T)(g) = \begin{cases} d_\theta \langle T\theta(k_1^{-1})v, \theta(k_2)v \rangle & \text{if } g = k_1 s k_2 \in KsK \\ 0 & \text{otherwise.} \end{cases}$$

Chapter 3
Multiplicity-Free Triples

This chapter is devoted to the study of multiplicity-free triples and their associated spherical functions. After the characterization of multiplicity-freenes in terms of commutativity of the associated Hecke algebra (Theorem 3.1), in Sect. 3.1 we present a generalization of a criterion due to Bump and Ginzburg [8]. In the subsequent section, we develop the intrinsic part of the theory of spherical functions, that is, we determine all their properties (e.g. the Functional Equation in Theorem 3.4) that may be deduced without their explicit form as matrix coefficients, as examined in Sect. 3.3. In Sect. 3.4 we consider the case when the K-representation (V, θ) is one-dimensional. This case is treated in full details in Chapter 13 of our monograph [17]. We refer to the CIMPA lecture notes by Faraut [31] for an excellent classical reference in the case of Gelfand pairs.

The results presented here will provide some of the necessary tools to present our main new examples of multiplicity-free triples (see Chaps. 5 and 6). Finally, in Sect. 3.6 we present a Frobenius–Schur type theorem for multiplicity-free triples: it provides a criterion for determining the type of a given irreducible spherical representation, namely, being real, quaternionic, or complex.

Let G and $K \leq G$ be finite groups. Let $\theta \in \widehat{K}$.

Theorem 3.1 *The following conditions are equivalent:*

(a) *The induced representation* $\mathrm{Ind}_K^G \theta$ *is multiplicity-free;*
(b) *the Hecke algebra* $\mathscr{H}(G, K, \psi)$ *is commutative.*

Moreover, suppose that one of the above condition is satisfied and that (1.29) *is an explicit multiplicity-free decomposition of* $\mathrm{Ind}_K^G V$ *into irreducibles. For every* $f \in \mathscr{H}(G, K, \psi)$ *consider the restriction* $T'_f = T_f|_{\mathscr{I}(G,K,\psi)}$ *to the invariant subspace* $\mathscr{I}(G, K, \psi)$*. Then, recalling that* J *denotes a (fixed) complete set of*

T. Ceccherini-Silberstein et al., *Gelfand Triples and Their Hecke Algebras*,
Lecture Notes in Mathematics 2267, https://doi.org/10.1007/978-3-030-51607-9_3

pairwise inequivalent irreducible G-representations contained in $\mathrm{Ind}_K^G\theta$, *we have that*

$$\mathscr{I}(G, K, \psi) = \bigoplus_{\sigma\in J} T_v\,[T_\sigma W_\sigma] \tag{3.1}$$

is the decomposition of $\mathscr{I}(G, K, \psi)$ *into* T'_f*-eigenspaces (cf. (4) in Theorem 2.2). Also, if* $\lambda_{\sigma,f}$ *is the eigenvalue of* T'_f *associated with the subspace* $T_v\,[T_\sigma W_\sigma]$, *then the map*

$$f \mapsto (\lambda_{\sigma,f})_{\sigma\in J} \tag{3.2}$$

is an algebra isomorphism from $\mathscr{H}(G, K, \psi)$ *onto* $\mathbb{C}^J$.

Finally, with the notation of Corollary 2.1, we have

$$\dim \mathscr{H}(G, K, \psi) = |J| = \sum_{s\in\mathscr{S}} \dim \mathrm{Hom}_{K_s}(\mathrm{Res}^K_{K_s}\theta, \theta^s). \tag{3.3}$$

Proof The Hecke algebra $\mathscr{H}(G, K, \psi)$ is antiisomorphic to $\mathrm{End}_G(\mathrm{Ind}_K^G V)$ via the map $\xi \circ S_v^{-1}$ by virtue of Theorems 2.1 and 2.2(3). Then the equivalence (a) $\Leftrightarrow$ (b) follows from the isomorphism (1.28).

By Theorem 2.2.(4), T'_f intertwines each irreducible representation $T_v\,[T_\sigma W_\sigma]$ with itself and therefore, by Schur's lemma, $T_f|_{T_v[T_\sigma W_\sigma]} = T'_f|_{T_v[T_\sigma W_\sigma]}$ is a multiple of the identity on this space.

Suppose that $f_1 \in \mathscr{I}(G, K, \psi)$ and $f_1 = \sum_{\sigma\in J} f_\sigma$ with $f_\sigma \in T_v\,[T_\sigma W_\sigma]$. Then $T_f(f_1) = \sum_{\sigma\in J} \lambda_{\sigma,f} f_\sigma$ proving the required properties for the map (3.2) (cf. (1.8)). In particular, $\dim \mathscr{H}(G, K, \psi) = \dim \mathbb{C}^J = |J|$ (see also Proposition 1.3). Since $\dim \mathscr{H}(G, K, \psi) = \dim \widetilde{\mathscr{H}}(G, K, \psi)$, the second equality in (3.3) follows from Corollary 2.1. □

Definition 3.1 If one of the equivalent conditions (a) and (b) in Theorem 3.1 is satisfied, we say that (G, K, θ) is a *multiplicity-free triple*.

Observe that if $\theta = \iota_K$ is the trivial K-representation, then (G, K, ι_K) is a multiplicity-free triple if and only if (G, K) is a Gelfand pair (cf. Definition 1.1 and Theorem 1.1). More generally, for $\dim\theta = 1$ we also refer to [46, 64, 68], and [49, 50].

The following proposition provides a useful tool for proving multiplicity-freeness for certain triples. The map $T \mapsto T^\sharp$ therein is a fixed antiautomorphism of the algebra $\mathrm{End}(V)$ (for instance, adjunction, or transposition). Recall (2.24) for the definition of ψ and its relation to θ.

Proposition 3.1 *Suppose there exists an antiautomorphism* τ *of* G *such that* $f(\tau(g)) = f(g)$ *for all* $f \in \mathscr{H}(G, K, \psi)$ *and* $g \in G$. *Then* $\mathscr{H}(G, K, \psi)$ *is commutative. Similarly, if* $F(\tau(g)) = F(g)^\sharp$ *for all* $F \in \widetilde{\mathscr{H}}(G, K, \theta)$ *and* $g \in G$, *then* $\widetilde{\mathscr{H}}(G, K, \theta)$ *is commutative.*

Proof Let $f_1, f_2 \in \mathscr{H}(G, K, \psi)$ and $g \in G$. We have

$$
\begin{aligned}
[f_1 * f_2](g) &= \sum_{h \in G} f_1(gh) f_2(h^{-1}) \\
(f \circ \tau = f) \quad &= \sum_{h \in G} f_1(\tau(gh)) f_2(\tau(h^{-1})) \\
&= \sum_{h \in G} f_1(\tau(h)\tau(g)) f_2(\tau(h^{-1})) \\
&= \sum_{h \in G} f_2(\tau(h^{-1})) f_1(\tau(h)\tau(g)) \\
(\text{setting } t = \tau(h)) \quad &= \sum_{t \in G} f_2(t^{-1}) f_1(t\tau(g)) \\
&= [f_2 * f_1](\tau(g)) = [f_2 * f_1](g),
\end{aligned}
$$

showing that $\mathscr{H}(G, K, \psi)$ is commutative. Similarly, for $F_1, F_2 \in \widetilde{\mathscr{H}}(G, K, \theta)$ and $g \in G$, we have

$$
\begin{aligned}
[F_1 * F_2](g) &= \sum_{h \in G} F_1(h^{-1}g) F_2(h) \\
(F \circ \tau = F^{\sharp}) \quad &= \sum_{h \in G} F_1(\tau(g)\tau(h)^{-1})^{\sharp} F_2(\tau(h))^{\sharp} \\
&= \sum_{h \in G} [F_2(\tau(h)) F_1(\tau(g)\tau(h^{-1}))]^{\sharp} \\
(\text{setting } \tau(g)\tau(h)^{-1} = t) \quad &= \left[\sum_{t \in G} F_2(t^{-1}\tau(g)) F_1(t) \right]^{\sharp} \\
&= [F_2 * F_1](\tau(g))^{\sharp} = [F_2 * F_1](g),
\end{aligned}
$$

showing that $\widetilde{\mathscr{H}}(G, K, \theta)$ is commutative as well. □

If the first condition in the above proposition is satisfied, we say that the multiplicity-free triple (G, K, θ) is *weakly symmetric*. When $\tau(g) = g^{-1}$ we say that (G, K, θ) is *symmetric*.

Example 3.1 (Weakly Symmetric Gelfand Pairs) Suppose that $\theta = \iota_K$ is the trivial representation of K. Then $\mathscr{H}(G, K, \psi) = {}^K L(G)^K$ and condition $f(g^{-1}) = f(g)$ (resp. $f(\tau(g)) = f(g)$), for all $f \in \mathscr{H}(G, K, \psi)$ and $g \in G$, is equivalent to $g^{-1} \in KgK$ (resp. $\tau(g) \in KgK$), for all $g \in G$. If this holds, one says that (G, K) is a *symmetric* (resp. *weakly symmetric*) Gelfand pair (cf. [11, Example 4.3.2 and Exercise 4.3.3]).

3.1 A Generalized Bump–Ginzburg Criterion

We now present a generalization of a criterion due to Bump and Ginzburg [8] (see also [17, Corollary 13.3.6.(a)]). We use the notation in (2.9) and we assume that $T \mapsto T^\sharp$ is a fixed antiautomorphism of the algebra $\mathrm{End}(V)$.

Theorem 3.2 (A Generalized Bump–Ginzburg Criterion) *Let G be a finite group, $K \leq G$ a subgroup, and $(\theta, V) \in \widehat{K}$. Suppose that there exists an antiautomorphism $\tau : G \to G$ such that:*

- *K is τ-invariant: $\tau(K) = K$,*
- *$\theta(\tau(k)) = \theta(k)^\sharp$ for all $k \in K$,*
- *for every $s \in \mathscr{S}_0$ (cf. (2.10)) there exist $k_1, k_2 \in K$ such that*

$$\tau(s) = k_1 s k_2 \tag{3.4}$$

and

$$\theta(k_2)^* T \theta(k_1)^* = T^\sharp \tag{3.5}$$

for all $T \in \mathrm{Hom}_{K_s}(\mathrm{Res}^K_{K_s}\theta, \theta^s)$,

- *$(T^\sharp)^* = (T^*)^\sharp$ for all $T \in \mathrm{End}(V)$ (i.e., $\sharp$ and $*$ commute).*

Then the triple (G, K, θ) is multiplicity-free.

Proof Let $s \in \mathscr{S}_0$. We show that $\mathscr{L}_T(\tau(g)) = \mathscr{L}_T(g)^\sharp$ for all $g \in G$ and $T \in \mathrm{Hom}_{K_s}(\mathrm{Res}^K_{K_s}\theta, \theta^s)$ (cf. (2.9)), so that we may apply the second condition in Proposition 3.1. First of all, we have

$$\mathscr{L}_T(\tau(s)) = \mathscr{L}_T(k_1 s k_2) = \theta(k_2)^* T \theta(k_1)^* = T^\sharp = \mathscr{L}_T(s)^\sharp \tag{3.6}$$

Let $g \in G$ and suppose that $g = h_1 s h_2$ with $s \in \mathscr{S}_0$ and $h_1, h_2 \in K$. Then

$$\begin{aligned}
\mathscr{L}_T(\tau(g)) &= \mathscr{L}_T(\tau(h_2)\tau(s)\tau(h_1)) \\
&= \theta(\tau(h_1))^* \mathscr{L}_T(\tau(s)) \theta(\tau(h_2))^* \\
\text{(by (3.6))} &= (\theta(h_1)^\sharp)^* T^\sharp (\theta(h_2)^\sharp)^* \\
&= (\theta(h_1)^*)^\sharp T^\sharp (\theta(h_2)^*)^\sharp \\
&= [\theta(h_2)^* \mathscr{L}_T(s) \theta(h_1)^*]^\sharp \\
&= \mathscr{L}_T(g)^\sharp.
\end{aligned}$$

□

Remark 3.1 The Bump–Ginzburg criterion, that concerns the case $\dim\theta = 1$ so that $\mathrm{End}(V_\theta) \cong \mathbb{C}$ is commutative, may be obtained by taking the operator $\sharp$ as the identity. As in the following we shall only make use of this version, rather than its generalization (Theorem 3.2), for the convenience of the reader we state it as in its original form.

Theorem 3.3 (Bump–Ginzburg Criterion) *Let G be a finite group, $K \leq G$ a subgroup, and χ a one-dimensional K-representation. Suppose that there exists an antiautomorphism $\tau : G \to G$ such that:*

- *K is τ-invariant: $\tau(K) = K$,*
- *$\chi(\tau(k)) = \chi(k)$ for all $k \in K$,*
- *for every $s \in \mathcal{S}_0$ (cf. (2.10)) there exist $k_1, k_2 \in K$ such that*

$$\tau(s) = k_1 s k_2 \tag{3.7}$$

and

$$\chi(k_1)\chi(k_2) = 1. \tag{3.8}$$

Then the triple (G, K, χ) is multiplicity-free.

Open Problem In [15, Section 4], there is a quite deep analysis of the Mackey-Gelfand criterion for finite Gelfand pairs. It should be interesting to extend that analysis in the present setting. Also, the twisted Frobenius–Schur theorem (cf. [15, Section 9]) deserves to be analyzed in the present setting (cf. Sect. 3.6).

3.2 Spherical Functions: Intrinsic Theory

In this section we develop the intrinsic part of the theory of spherical functions, that is, we determine all their properties that may be deduced without their explicit form as matrix coefficients.

Suppose now that (G, K, θ) is a multiplicity-free triple.

Definition 3.2 A function $\phi \in \mathcal{H}(G, K, \psi)$ is *spherical* provided it satisfies the following conditions:

$$\phi(1_G) = 1 \tag{3.9}$$

and, for all $f \in \mathcal{H}(G, K, \psi)$, there exists $\lambda_{\phi,f} \in \mathbb{C}$ such that

$$\phi * f = \lambda_{\phi,f}\phi. \tag{3.10}$$

We denote by $\mathcal{S}(G, K, \psi) \subseteq \mathcal{H}(G, K, \psi)$ the set of all spherical functions associated with the multiplicity-free triple (G, K, θ).

Condition (3.10) may be reformulated by saying that ϕ is an eigenvector for the convolution operator T_f for every $f \in \mathcal{H}(G, K, \psi)$. Moreover, combining (3.10) and (3.9), we get $\lambda_{\phi,f} = [\phi * f](1_G)$, and, taking into account the commutativity of $\mathscr{H}(G, K, \psi)$, we deduce the following equivalent reformulation of (3.10):

$$\phi * f = f * \phi = ([\phi * f](1_G))\,\phi. \tag{3.11}$$

We now give a basic characterization of the spherical functions that makes use of the function $\psi \in L(G)$ defined in (2.24).

Theorem 3.4 (Functional Equation) *A non-trivial function $\phi \in L(G)$ is spherical if and only if*

$$\sum_{k\in K} \phi(gkh)\overline{\psi(k)} = \phi(g)\phi(h), \qquad \textit{for all } g, h \in G. \tag{3.12}$$

Proof Suppose that $0 \neq \phi \in L(G)$ satisfies (3.12). Choose $h \in G$ such that $\phi(h) \neq 0$. Then writing (3.12) in the form $\phi(g) = \frac{1}{\phi(h)} \sum_{k\in K} \phi(gkh)\overline{\psi(k)}$ we get

$$\begin{aligned}
[\phi * \psi](g) &= \frac{1}{\phi(h)} \sum_{k,k_1\in K} \phi(gk_1kh)\overline{\psi(k)}\psi(k_1^{-1})\\
(k_1k = k_2) \quad &= \frac{1}{\phi(h)} \sum_{k_2\in K} \phi(gk_2h)\overline{[\psi * \psi](k_2)}\\
(\text{by } (2.26)) \quad &= \frac{1}{\phi(h)} \sum_{k_2\in K} \phi(gk_2h)\overline{\psi(k_2)}\\
&= \phi(g)
\end{aligned}$$

for all $g \in G$, that is, $\phi * \psi = \phi$. Similarly one proves that $\psi * \phi = \phi$. Combining these two facts together we get $\psi * \phi * \psi = \phi$ yielding $\phi \in \mathscr{H}(G, K, \psi)$. Then setting $h = 1_G$ in (3.12) we get

$$\phi(g)\phi(1_G) = \sum_{k\in K} \phi(gk)\overline{\psi(k)} = [\phi * \psi](g) = \phi(g)$$

for all $g \in G$, and therefore (3.9) is satisfied. Finally, for $f \in \mathscr{H}(G, K, \psi)$ and $g \in G$ we have

$$\begin{aligned}
[\phi * f](g) &= [\phi * f * \psi](g)\\
&= \sum_{h\in G} \sum_{k\in K} \phi(gkh) f(h^{-1})\overline{\psi(k)}\\
(\text{by } (3.12)) \quad &= \phi(g) \sum_{h\in G} \phi(h) f(h^{-1})\\
&= [\phi * f](1_G)\phi(g)
\end{aligned}$$

and also (3.11) is satisfied. This shows that ϕ is spherical.

Conversely, suppose that ϕ is spherical and set

$$F_g(h) = \sum_{k \in K} \phi(gkh)\overline{\psi(k)},$$

for all $h, g \in G$. If $f \in \mathscr{H}(G, K, \psi)$ and $g, g_1 \in G$, then we have

$$\begin{aligned}
[F_g * f](g_1) &= \sum_{k \in K} \sum_{h \in G} \phi(gkg_1h) f(h^{-1})\overline{\psi(k)} \\
\text{(by (3.11))} \quad &= [\phi * f](1_G) \sum_{k \in K} \phi(gkg_1)\overline{\psi(k)} \\
&= [\phi * f](1_G) F_g(g_1).
\end{aligned} \tag{3.13}$$

Let now show that the function $J_g \in L(G)$, defined by

$$J_g(h) = \sum_{k \in K} f(hkg)\overline{\psi(k)}$$

for all $h \in G$, is in $\mathscr{H}(G, K, \psi)$ for every $g \in G$. Indeed, for all $h \in G$,

$$\begin{aligned}
[\psi * J_g * \psi](h) &= \sum_{k, k_1, k_2 \in K} \psi(k_1) f(k_1^{-1} h k_2^{-1} kg) \psi(k_2)\overline{\psi(k)} \\
(k_3 = k_2^{-1}k) \quad &= \sum_{k, k_3 \in K} [\psi * f](hk_3g)\psi(kk_3^{-1})\psi(k^{-1}) \\
&= \sum_{k_3 \in K} f(hk_3g)[\psi * \psi](k_3^{-1}) \\
\text{(by (2.26))} \quad &= \sum_{k_3 \in K} f(hk_3g)\overline{\psi(k_3)} \\
&= J_g(h).
\end{aligned}$$

Moreover,

$$\begin{aligned}
[\phi * J_g](1_G) &= \sum_{h \in G} \phi(h^{-1}) \sum_{k \in K} f(hkg)\overline{\psi(k)} \\
(hk = t) \quad &= \sum_{t \in G} \left(\sum_{k \in K} \psi(k^{-1})\phi(kt^{-1}) \right) f(tg) \\
&= \sum_{t \in G} \phi(t^{-1}) f(tg) \\
&= [\phi * f](g).
\end{aligned} \tag{3.14}$$

It follows that, for $g, g_1 \in G$,

$$\begin{aligned}
[F_g * f](g_1) &= \sum_{h \in G} \sum_{k \in K} \phi(gkg_1h)\overline{\psi(k)} f(h^{-1}) \\
(kg_1h = t) \quad &= \sum_{t \in G} \phi(gt) \sum_{k \in K} \overline{\psi(k)} f(t^{-1}kg_1) \\
&= \phi * J_{g_1}(g) \\
(\text{by } (3.11)) \quad &= [\phi * J_{g_1}](1_G)\phi(g) \\
(\text{by } (3.14)) \quad &= [\phi * f](g_1)\phi(g) \\
(\text{again by } (3.11)) \quad &= [\phi * f](1_G)\phi(g_1)\phi(g).
\end{aligned} \tag{3.15}$$

From (3.13) and (3.15) we get $[\phi * f](1_G)F_g(g_1) = [\phi * f](1_G)\phi(g_1)\phi(g)$ and taking $f \in \mathscr{H}(G, K, \psi)$ such that $[\phi * f](1_G) \neq 0$ (take, for instance, $f = \psi * \delta_{1_G} * \psi$) this yields $F_g(g_1) = \phi(g_1)\phi(g)$ which is nothing but (3.12). □

Remark 3.2 In the case $\psi = 1$, (3.12) yields the well known product formula for spherical functions. There is an abstract "addition theorem" which is the expansion of the map $k \mapsto \phi(gkh)$ as a sum of functions of different K-isotype. The product formula is the "integrated" version.

A linear functional $\Phi : \mathscr{H}(G, K, \psi) \to \mathbb{C}$ is said to be *multiplicative* provided that

$$\Phi(f_1 * f_2) = \Phi(f_1)\Phi(f_2)$$

for all $f_1, f_2 \in \mathscr{H}(G, K, \psi)$.

Theorem 3.5 *Let ϕ be a spherical function and set*

$$\Phi(f) = [f * \phi](1_G) \tag{3.16}$$

for all $f \in \mathscr{H}(G, K, \psi)$. Then Φ is a linear multiplicative functional on $\mathscr{H}(G, K, \psi)$. Conversely, any non-trivial multiplicative linear functional on $\mathscr{H}(G, K, \psi)$ is of this form.

Proof Let $f_1, f_2 \in \mathscr{H}(G, K, \psi)$. Then, we get:

$$\begin{aligned}
\Phi(f_1 * f_2) &= [(f_1 * f_2) * \phi](1_G) \\
&= [f_1 * (f_2 * \phi)](1_G) \\
(\text{by } (3.11)) \quad &= [f_1 * ([f_2 * \phi](1_G)\phi)]\,(1_G) \\
&= [f_1 * \phi](1_G)[f_2 * \phi](1_G) \\
&= \Phi(f_1)\Phi(f_2).
\end{aligned}$$

Conversely, suppose that Φ is a non-trivial multiplicative linear functional on $\mathscr{H}(G, K, \psi)$. We can extend Φ to a linear functional $\overline{\Phi}$ on $L(G)$ by setting $\overline{\Phi}(f) = \Phi(\psi * f * \psi)$ for all $f \in L(G)$. By Riesz' theorem, we can find $\varphi \in L(G)$ such that

$$\Phi(\psi * f * \psi) = \sum_{g \in G} f(g)\varphi(g^{-1}) = [f * \varphi](1_G),$$

for all $f \in L(G)$. From (2.33) we deduce that if $f_1 \in \mathscr{H}(G, K, \psi)$ then

$$\Phi(f_1) = [f_1 * \varphi](1_G) = [f_1 * \psi * \varphi * \psi](1_G)$$

and therefore the function $\varphi \in L(G)$ may be replaced by the function $\phi = \psi * \varphi * \psi \in \mathscr{H}(G, K, \psi)$. With this position, (2.33) also yields

$$\Phi(\psi * f * \psi) = [\phi * \psi * f * \psi](1_G) = [\phi * f](1_G) \tag{3.17}$$

for all $f \in L(G)$.

We are only left to show that $\phi \in \mathscr{S}(G, K, \psi)$. Let then $f_1 \in \mathscr{H}(G, K, \psi)$ and $f_2 \in L(G)$. Since Φ is multiplicative, the quantity

$$\begin{aligned} \Phi(f_1 * \psi * f_2 * \psi) &= [\phi * f_1 * \psi * f_2 * \psi](1_G) \\ \text{(by (2.33))} \quad &= [\phi * f_1 * f_2](1_G) \\ &= \sum_{h \in G} [\phi * f_1](h) f_2(h^{-1}) \end{aligned}$$

must be equal to

$$\Phi(f_1)\Phi(\psi * f_2 * \psi) = \Phi(f_1)[\phi * f_2](1_G) = \sum_{h \in G} \Phi(f_1)\phi(h) f_2(h^{-1}),$$

where the first equality follows from (3.17). Since $f_2 \in L(G)$ was arbitrary, we get the equality $[\phi * f_1](h) = \Phi(f_1)\phi(h) = [f_1 * \phi](1_G)\phi(h) = [\phi * f_1](1_G)\phi(h)$. Thus ϕ satisfies condition (3.11). Moreover, taking $h = 1_G$, one also obtains $\phi(1_G) = 1$, and (3.9) is also satisfied. In conclusion, ϕ is a spherical function. □

Corollary 3.1 *The number $|\mathscr{S}(G, K, \psi)|$ of spherical functions in $\mathscr{H}(G, K, \psi)$ equals the number $|J|$ of pairwise inequivalent irreducible G-representations contained in $\mathrm{Ind}_K^G \theta$.*

Proof We have $\mathscr{H}(G, K, \psi) \cong \mathbb{C}^J$ (see Theorem 3.1) and every linear multiplicative functional on $\mathbb{C}^J$ is of the form $\lambda = (\lambda_\sigma)_{\sigma \in J} \mapsto \lambda_{\overline{\sigma}}$, for a fixed $\overline{\sigma} \in J$. □

Proposition 3.2 *Let ϕ and $\phi_1 \neq \phi_2$ be spherical functions. Then*

(i) $\phi^* = \phi$;
(ii) $\phi_1 * \phi_2 = 0$;
(iii) $\langle \lambda_G(g_1)\phi_1, \lambda_G(g_2)\phi_2 \rangle_{L(G)} = 0$ *for all* $g_1, g_2 \in G$; *in particular* $\langle \phi_1, \phi_2 \rangle_{L(G)} = 0$.

Proof

(i) By definition of a spherical function, we have

$$\phi^* * \phi = [\phi^* * \phi](1_G)\phi = \left(\sum_{g \in G} \overline{\phi(g^{-1})}\phi(g^{-1}) \right) \phi = \|\phi\|^2 \phi.$$

On the other hand, since $(\phi^* * \phi)^* = \phi^* * \phi$, we have

$$[\phi^* * \phi](g) = \overline{[\phi^* * \phi](g^{-1})} = \overline{[\phi^* * \phi](1_G)} \cdot \overline{\phi(g^{-1})} = \|\phi\|^2 \overline{\phi(g^{-1})}$$

and therefore we must have $\phi = \phi^*$.

(ii) By commutativity of $\mathscr{H}(G, K, \psi)$, we have that

$$\begin{gathered}[\phi_1 * \phi_2](g) = ([\phi_1 * \phi_2](1_G))\, \phi_1(g) \\ \text{must be equal to} \\ [\phi_2 * \phi_1](g) = ([\phi_2 * \phi_1](1_G))\, \phi_2(g) = ([\phi_1 * \phi_2](1_G))\, \phi_2(g)\end{gathered} \tag{3.18}$$

for all $g \in G$. Therefore, if $\phi_1 \neq \phi_2$, (3.18) implies $[\phi_1 * \phi_2](1_G) = [\phi_2 * \phi_2](1_G) = 0$. But then, again, (3.18) yields $\phi_1 * \phi_2 = 0$.

(iii) Let $g_1, g_2 \in G$. Applying (i) and (ii) we have

$$\begin{aligned}\langle \lambda_G(g_1)\phi_1, \lambda_G(g_2)\phi_2 \rangle_{L(G)} &= \langle \phi_1, \lambda_G(g_1^{-1}g_2)\phi_2 \rangle_{L(G)} \\ &= [\phi_1 * \phi_2^*](g_1^{-1}g_2) = [\phi_1 * \phi_2](g_1^{-1}g_2) = 0.\end{aligned}$$

□

Corollary 3.2 *The spherical functions form an orthogonal basis for $\mathscr{H}(G, K, \psi)$.*

Proof This is an immediate consequence of the first equality in (3.3). □

In Theorem 3.7 we shall compute the orthogonality relations for the spherical functions.

Recall (cf. Theorem 2.2(a)) that $\mathscr{I}(G, K, \psi)$ is a subrepresentation of the left-regular representation of G. For $\phi \in \mathscr{S}(G, K, \psi)$ we denote by $U_\phi = \text{span}\{\lambda_G(g)\phi : g \in G\}$ the subspace of $L(G)$ generated by all G-translates of ϕ. Then the following holds.

Theorem 3.6 *For every $\phi \in \mathscr{S}(G, K, \psi)$ the space U_ϕ is an irreducible G-representation and*

$$\mathscr{I}(G, K, \psi) = \bigoplus_{\phi \in \mathscr{S}(G,K,\psi)} U_\phi \tag{3.19}$$

is the decomposition of $\mathscr{I}(G, K, \psi)$ into irreducibles.

Proof Let $\phi_1, \phi_2 \in \mathscr{S}(G, K, \psi)$. Then U_{ϕ_1} is G-invariant and contained in $\mathscr{I}(G, K, \psi)$. Moreover, by Proposition 3.2(iii), if $\phi_1 \neq \phi_2$, then the spaces U_{ϕ_1} and U_{ϕ_2} are orthogonal. Finally, we can invoke Corollary 3.1 and the fact that $\mathscr{I}(G, K, \psi)$ is multiplicity-free (cf. Theorems 2.2(1) and 3.1), to conclude that the direct sum in the Right Hand Side of (3.19) exhausts the whole of $\mathscr{I}(G, K, \psi)$ and that the U_ϕs are irreducible. □

The space U_ϕ is called the *spherical representation* associated with $\phi \in \mathscr{S}(G, K, \psi)$. In the next section different realizations of the spherical functions and of the representations are discussed.

3.3 Spherical Functions as Matrix Coefficients

Let (G, K, θ) be a multiplicity-free triple. Let also $(\sigma, W) \in \widehat{G}$ be a spherical representation (i.e. $\sigma \in J$, where J is as in Theorem 3.1).

By Frobenius reciprocity (cf. (1.26)), σ is contained in $\mathrm{Ind}_K^G\theta$ if and only if $\mathrm{Res}_K^G\sigma$ contains θ. Moreover, if this is the case, since (G, K, θ) is multiplicity-free, then the multiplicity of θ in $\mathrm{Res}_K^G\sigma$ must be exactly one.

This implies that there exists an *isometric* map $L_\sigma : V \to W$ such that $\mathrm{Hom}_K(V, \mathrm{Res}_K^G W) = \mathbb{C}L_\sigma$. Now we show that, by means of Frobenius reciprocity, the operator T_σ in (1.29) may be expressed in terms of L_σ; see [61] for more general results.

Proposition 3.3 *The map $T_\sigma : W \to \{f : G \to V\}$ defined by setting*

$$[T_\sigma w](g) = \sqrt{\frac{d_\sigma |K|}{d_\theta |G|}} L_\sigma^* \sigma(g^{-1}) w, \tag{3.20}$$

for all $g \in G$ and $w \in W$, is an isometric immersion of W into $\mathrm{Ind}_K^G V$.

Proof We first note that, by (1.3), $L_\sigma^* \in \mathrm{Hom}_K(\mathrm{Res}_K^G W, V)$, that is $L_\sigma^*\sigma(k) = \theta(k)L_\sigma^*$ for all $k \in K$. This implies that $T_\sigma w \in \mathrm{Ind}_K^G V$ since

$$[T_\sigma w](gk) = \sqrt{\frac{d_\sigma|K|}{d_\theta|G|}} L_\sigma^*\sigma(k^{-1}g^{-1})w$$

$$= \sqrt{\frac{d_\sigma|K|}{d_\theta|G|}} \theta(k^{-1})L_\sigma^*\sigma(g^{-1})w = \theta(k^{-1})[T_\sigma w](g)$$

for all $g \in G$ and $k \in K$. It is also easy to check that $T_\sigma\sigma(g) = \lambda(g)T_\sigma$ for all $g \in G$, in other words $T_\sigma \in \mathrm{Hom}_G(W, \mathrm{Ind}_K^G V)$.

It remains to show that T_σ is an isometric embedding. First of all, it is straightforward to check that the operator

$$\sum_{g\in G} \sigma(g)L_\sigma L_\sigma^*\sigma(g^{-1})$$

belongs to $\mathrm{End}_G(W)$ and therefore, by Schur's lemma, it is a multiple of the identity: there exists $\alpha \in \mathbb{C}$ such that $\sum_{g\in G} \sigma(g)L_\sigma L_\sigma^*\sigma(g^{-1}) = \alpha I_W$. Since $L_\sigma : V \to W$ is an isometric embedding, we have that $L_\sigma^* L_\sigma = I_V$, the identity map $V \to V$, yielding $\mathrm{tr}(L_\sigma^* L_\sigma) = d_\theta$, so that

$$\mathrm{tr}\left(\sum_{g\in G} \sigma(g)L_\sigma L_\sigma^*\sigma(g^{-1})\right) = |G|\mathrm{tr}\left(L_\sigma L_\sigma^*\right) = |G|\mathrm{tr}\left(L_\sigma^* L_\sigma\right) = |G|d_\theta.$$

Since $\mathrm{tr}(\alpha I_W) = \alpha d_\sigma$, we get $\alpha = \frac{|G|d_\theta}{d_\sigma}$, that is,

$$\sum_{g\in G} \sigma(g)L_\sigma L_\sigma^*\sigma(g^{-1}) = \frac{|G|d_\theta}{d_\sigma} I_W.$$

We deduce that

$$\langle T_\sigma w_1, T_\sigma w_2\rangle_{\mathrm{Ind}_K^G V} = \frac{d_\sigma}{d_\theta|G|} \sum_{g\in G} \left\langle L_\sigma^*\sigma(g^{-1})w_1, L_\sigma^*\sigma(g^{-1})w_2\right\rangle_V$$

$$= \frac{d_\sigma}{d_\theta|G|} \left\langle w_1, \sum_{g\in G} \sigma(g)L_\sigma L_\sigma^*\sigma(g^{-1})w_2\right\rangle_W$$

$$= \langle w_1, w_2\rangle_W$$

for all $w_1, w_2 \in W$, thus showing that T_σ is isometric. □

From now on, in order to emphasize the dependence of the representation space W on $\sigma \in J$, we denote the former by W_σ. Moreover, with the notation in Sect. 2.2 (cf. (2.24), so that $v = v_1$), for each $\sigma \in J$ we set

$$w^\sigma = L_\sigma v \in W_\sigma. \tag{3.21}$$

Lemma 3.1 *Let $\sigma \in J$. Consider the map $S_\sigma : W_\sigma \to L(G)$ defined by setting*

$$[S_\sigma w](g) = \sqrt{\frac{d_\sigma}{|G|}} \langle w, \sigma(g) w^\sigma \rangle_{W_\sigma}, \tag{3.22}$$

for all $w \in W_\sigma$ and $g \in G$. Then $S_\sigma = T_v T_\sigma$, that is, the following diagram

$$\begin{array}{ccc} W_\sigma & \xrightarrow{T_\sigma} & \mathrm{Ind}_K^G V \\ & \searrow^{S_\sigma} & \downarrow T_v \\ & & \mathscr{I}(G, K, \psi) \end{array}$$

is commutative. As a consequence, S_σ is an isometric immersion of W_σ into $\mathscr{I}(G, K, \psi)$. Moreover, (3.1) *may be written in the form*

$$\mathscr{I}(G, K, \psi) = \bigoplus_{\sigma \in J} S_\sigma W_\sigma. \tag{3.23}$$

Proof For all $w \in W_\sigma$ and $g \in G$, we have

$$\begin{aligned} (T_v T_\sigma w)(g) &= \sqrt{d_\theta/|K|} \langle [T_\sigma w](g), v \rangle_V && \text{(by (2.25))} \\ &= \sqrt{d_\sigma/|G|} \langle L_\sigma^* \sigma(g^{-1}) w, v \rangle_V && \text{(by (3.20))} \\ &= \sqrt{d_\sigma/|G|} \langle w, \sigma(g) L_\sigma v \rangle_{W_\sigma} \\ &= [S_\sigma w](g) && \text{(by (3.22)).} \end{aligned}$$

□

Remark 3.3 Clearly, ψ and $\mathscr{I}(G, K, \psi)$ depend on the choice of $v \in V$. From (3.22) and the orthogonality relations (1.14) it follows that the replacement of v with an orthogonal vector u leads to an isomorphic orthogonal realization of $\mathscr{I}(G, K, \psi)$.

Equations (3.19) and (3.23) provide two different decompositions of $\mathscr{I}(G, K, \psi)$ into irreducible representations: in the first one these are constructed in terms of spherical functions, in the second one they come from a decomposition of $\mathrm{Ind}_K^G V$. The connection between these two decompositions requires an explicit expression for the spherical functions. To this end, given $\sigma \in J$, we define $\phi^\sigma \in L(G)$ by setting

$$\phi^\sigma(g) = \langle w^\sigma, \sigma(g) w^\sigma \rangle_{W_\sigma}, \tag{3.24}$$

for all $g \in G$, where w^σ is as in (3.21).

Our next task is to show that the above defined functions ϕ^σ, $\sigma \in J$, are spherical and that, in fact, any spherical function is one of these. In other words, $\mathscr{S}(G, K, \psi) = \{\phi^\sigma : \sigma \in J\}$.

We need to prove a preliminary identity. We choose an orthonormal basis $\{u_i : i = 1, 2, \ldots, d_\sigma\}$ for W_σ in the following way. Let $\mathrm{Res}^G_K W_\sigma = L_\sigma V \oplus (\oplus_\eta m_\eta U_\eta)$ an explicit (cf. Remark 1.1) decomposition of $\mathrm{Res}^G_K W_\sigma$ into irreducible K-representations (the U_η's are pairwise distinct and each of them is non-equivalent to V; m_η is the multiplicity of U_η). We set $u_1 = w^\sigma = L_\sigma v$, and let $\{u_i : 1 \leq i \leq d_\theta\}$ form an orthonormal basis for $L_\sigma V$. Finally, the remaining u_i's are only subject to the condition of belonging to some irreducible U_η of the explicit decomposition. Then, by (1.14), we have

$$\sum_{k \in K} \langle u_1, \sigma(k) u_1 \rangle \langle \sigma(k) u_i, u_j \rangle = \frac{|K|}{d_\theta} \delta_{1i} \delta_{1j}.$$

Since $\psi(k) = \frac{d_\theta}{|K|} \langle u_1, \sigma(k) u_1 \rangle_{W_\sigma}$, the above may be written in the form

$$\left\langle \sum_{k \in K} \psi(k) \sigma(k) u_i, u_j \right\rangle = \delta_{1i} \delta_{1j}$$

yielding

$$\sum_{k \in K} \psi(k) \sigma(k) u_i = \delta_{i1} w^\sigma. \tag{3.25}$$

Theorem 3.7 *$\phi^\sigma \in L(G)$ is the spherical function associated with W_σ, that is, in the notation of Theorem 3.6 and Lemma 3.1, $U_{\phi^\sigma} = S_\sigma W_\sigma$. Moreover, the following orthogonality relations hold:*

$$\langle \phi^\sigma, \phi^\rho \rangle_{L(G)} = \frac{|G|}{d_\sigma} \delta_{\sigma,\rho}, \tag{3.26}$$

for all $\sigma, \rho \in J$.

Proof By (3.22) we have $\phi^\sigma = \sqrt{\frac{|G|}{d_\sigma}} S_\sigma w^\sigma$ and therefore, by Lemma 3.1, ϕ^σ belongs to the subspace of $\mathscr{I}(G, K, \psi)$ isomorphic to W_σ, namely to $S_\sigma W_\sigma$. Now we check the functional identity (3.12) to show that ϕ^σ is a spherical function: for all $g, h \in G$,

$$\begin{aligned}\sum_{k\in K}\phi^\sigma(gkh)\overline{\psi(k)} &= \sum_{k\in K}\langle w^\sigma, \sigma(gkh)w^\sigma\rangle\overline{\psi(k)}\\ &= \sum_{i=1}^{d_\sigma}\langle\sigma(g^{-1})w^\sigma, u_i\rangle\sum_{k\in K}\overline{\langle\sigma(kh)w^\sigma, u_i\rangle\psi(k)}\\ &= \sum_{i=1}^{d_\sigma}\langle\sigma(g^{-1})w^\sigma, u_i\rangle\overline{\left\langle\sigma(h)w^\sigma, \sum_{k\in K}\psi(k^{-1})\sigma(k^{-1})u_i\right\rangle}\\ \text{(by (3.25))} &= \phi^\sigma(g)\phi^\sigma(h).\end{aligned}$$

Finally, (3.26) is a particular case of (1.14). □

The *spherical Fourier transform* is the map

$$\mathcal{F}: \mathcal{H}(G, K, \psi) \longrightarrow L(J) \equiv \mathbb{C}^J$$

defined by setting

$$[\mathcal{F}f](\sigma) = \langle f, \phi^\sigma\rangle = \sum_{g\in G} f(g)\overline{\phi^\sigma(g)}$$

for all $f \in \mathcal{H}(G, K, \psi)$ and $\sigma \in J$. Note that

$$[\mathcal{F}f](\sigma) = \sum_{g\in G} f(g)\overline{\phi^\sigma(g)} = \sum_{g\in G} f(g)\phi^\sigma(g^{-1}) = [f * \phi^\sigma](1_G). \tag{3.27}$$

From Corollary 3.2 and the orthogonality relations (3.26) we immediately deduce the *inversion formula*:

$$f = \frac{1}{|G|}\sum_{\sigma\in J} d_\sigma[\mathcal{F}f](\sigma)\phi^\sigma \tag{3.28}$$

and the *Plancherel formula*:

$$\langle f_1, f_2\rangle_{L(G)} = \frac{1}{|G|}\sum_{\sigma\in J} d_\sigma[\mathcal{F}f_1](\sigma)\overline{[\mathcal{F}f_2](\sigma)},$$

for all $f, f_1, f_2 \in \mathcal{H}(G, K, \psi)$. Moreover, the *convolution formula*

$$\mathcal{F}(f_1 * f_2) = (\mathcal{F}f_1)(\mathcal{F}f_2)$$

follows from the inversion formula and (1.15).

Proposition 3.4 *Let $f \in \mathscr{H}(G, K, \psi)$. Then the eigenvalue of $T_f \in \mathrm{End}_G(\mathscr{I}(G, K, \psi))$ (see Theorem 2.2.(4)) associated with the eigenspace $S_\sigma W_\sigma$ (see (3.23)) is equal to $[\mathscr{F} f](\sigma)$, for all $\sigma \in J$.*

Proof This is a simple calculation:

$$\begin{aligned}[T_f\phi^\sigma](g) &= [f * \phi^\sigma](g)\\ \text{(by (3.11))} \quad &= [f * \phi^\sigma](1_G)\phi^\sigma(g)\\ \text{(by (3.27))} \quad &= [\mathscr{F} f](\sigma)\phi^\sigma(g).\end{aligned}$$

□

Proposition 3.5 *The operator $E_\sigma : \mathscr{I}(G, K, \psi) \longrightarrow L(G)$, defined by setting*

$$E_\sigma f = \frac{d_\sigma}{|G|}[f * \phi^\sigma]$$

for all $f \in \mathscr{I}(G, K, \psi)$, is the orthogonal projection from $\mathscr{I}(G, K, \psi)$ onto $S_\sigma W_\sigma$.

Proof First of all note that, for $g \in G$ and $f \in \mathscr{I}(G, K, \psi)$, we have:

$$\begin{aligned}[E_\sigma f](g) &= \frac{d_\sigma}{|G|}\sum_{h\in G} f(h)\phi^\sigma(h^{-1}g) = \frac{d_\sigma}{|G|}\sum_{h\in G} f(h)\overline{\phi^\sigma(g^{-1}h)}\\ &= \frac{d_\sigma}{|G|}\langle f, \lambda_G(g)\phi^\sigma\rangle.\end{aligned}$$

Therefore, for $\eta \neq \sigma$ and $h \in G$,

$$\left[E_\sigma \lambda_G(h)\phi^\eta\right](g) = \frac{d_\sigma}{|G|}\langle \lambda_G(h)\phi^\eta, \lambda_G(g)\phi^\sigma\rangle = 0$$

by Proposition 3.2(iii), that is, $\bigoplus_{\eta\in J, \eta\neq\sigma} S_\eta W_\eta \subseteq \ker E_\sigma$. Similarly,

$$\begin{aligned}\left[E_\sigma \lambda_G(h)\phi^\sigma\right](g) &= \frac{d_\sigma}{|G|}\langle \lambda_G(h)\phi^\sigma, \lambda_G(g)\phi^\sigma\rangle\\ &= \frac{d_\sigma}{|G|}\langle \phi^\sigma, \lambda_G(h^{-1}g)\phi^\sigma\rangle\\ &= \frac{d_\sigma}{|G|}[\phi^\sigma * \phi^\sigma](h^{-1}g)\\ \text{(by (3.11))} \quad &= \frac{d_\sigma}{|G|}[\phi^\sigma * \phi^\sigma](1_G)\phi^\sigma(h^{-1}g)\\ \text{(since } [\phi^\sigma * \phi^\sigma](1_G) = \|\phi^\sigma\|^2 = |G|/d_\sigma) \quad &= \lambda_G(h)\phi^\sigma(g).\end{aligned}$$

□

We now prove the relations between the spherical function ϕ^σ and the character χ^σ of σ.

Proposition 3.6 *For all $g \in G$ we have:*

$$\chi^\sigma(g) = \frac{d_\sigma}{|G|} \sum_{h \in G} \overline{\phi^\sigma(h^{-1}gh)} \tag{3.29}$$

and

$$\phi^\sigma(g) = [\overline{\chi^\sigma} * \psi](g). \tag{3.30}$$

Proof Clearly, (3.29) is just a particular case of (1.17), keeping into account the definition of ϕ^σ (see (3.24)). Using the bases in (3.25) we have

$$\begin{aligned}[\overline{\chi^\sigma} * \psi](g) &= \sum_{k \in K} \sum_{i=1}^{d_\sigma} \overline{\langle \sigma(gk^{-1})u_i, u_i \rangle} \psi(k) \\ &= \sum_{k \in K} \sum_{i=1}^{d_\sigma} \overline{\langle \psi(k^{-1})\sigma(k^{-1})u_i, \sigma(g^{-1})u_i \rangle} \\ \text{(by (3.25))} &= \phi^\sigma(g).\end{aligned}$$

□

3.4 The Case dim $\theta = 1$

In this section we consider the case when the K-representation (V, θ) is one-dimensional (this case is also treated, in a more detailed way, in [17, Chapter 13]). We then denote by $\chi = \chi^\theta$ its character. Let $\mathscr{S}_0 \subseteq \mathscr{S} \subseteq G$ and K_s, $s \in \mathscr{S}$, be as in Sect. 2.1, and let $\psi \in L(G)$ and $\mathscr{H}(G, K, \psi)$ be as in Sect. 2.2.

Note that, in our setting, we have (cf. (2.24)) $\psi(k) = \frac{1}{|K|}\overline{\chi(k)}$ for all $k \in K$.

Theorem 3.8

(1) $\mathscr{H}(G, K, \psi) = \{f \in L(G) : f(k_1 g k_2) = \overline{\chi(k_1)\chi(k_2)} f(g)$ *for all* $k_1, k_2 \in K$ *and* $g \in G\}$;
(2) $\mathscr{S}_0 = \{s \in \mathscr{S} : \chi(s^{-1}xs) = \chi(x)$, *for all* $x \in K_s\}$;
(3) *every function* $f \in \mathscr{H}(G, K, \psi)$ *only depends on its values on* $\mathscr{S}_0$, *namely*

$$f(g) = \begin{cases} \overline{\chi(k_1)} f(s) \overline{\chi(k_2)} & \text{if } g = k_1 s k_2 \text{ with } s \in \mathscr{S}_0 \\ 0 & \text{otherwise.} \end{cases}$$

Proof

(1) Let $f \in \mathscr{H}(G, K, \psi)$. Then by Theorem 2.2(3) we can find a unique function $F \in \widetilde{\mathscr{H}}(G, K, \theta)$ such that $f = S_v(F)$. Since V is one-dimensional, we have $F \in L(G)$ and $\theta = \chi$ so that (2.1) yields

$$F(k_1 g k_2) = \overline{\chi(k_2)} F(g) \overline{\chi}(k_1) = \overline{\chi(k_1)\chi(k_2)} F(g),$$

which, by linearity of S_v, yields

$$f(k_1 g k_2) = \overline{\chi(k_1)\chi(k_2)} f(g)$$

for all $k_1, k_2 \in K$ and $g \in G$. This shows the inclusion $\subseteq$. Since S_v is bijective, (1) follows.

(2) We first observe that since θ is one-dimensional, so are $\mathrm{Res}^K_{K_s} \theta$ and θ^s for all $s \in \mathscr{S}$. As a consequence, for $s \in \mathscr{S}$ we have that $\mathrm{Hom}_{K_s}(\mathrm{Res}^K_{K_s} \theta, \theta^s)$ is non-trivial if and only if $\mathrm{Res}^K_{K_s} \theta$ equals θ^s and this is in turn equivalent to $\chi(x) = \chi(s^{-1} x s)$ for all $x \in K_s$.

(3) This follows immediately from (1), (2), and the fact that any $f \in \mathscr{H}(G, K, \psi)$ is supported in $\sqcup_{s \in \mathscr{S}_0} K s K$ (cf. Theorem 2.2(3)). □

3.5 An Example: The Gelfand–Graev Representation of GL(2, $\mathbb{F}_q$)

We now illustrate a fundamental example, which is completely examined in [17, Chapter 14]. This gives us the opportunity to introduce some notation and basic notions that shall be widely used in Chaps. 5 and 6.

Let p be a prime number, n a positive integer, and denote by $\mathbb{F}_q$ the field with $q := p^n$ elements. Let $G = \mathrm{GL}(2, \mathbb{F}_q)$ denote the *general linear group of rank 2 over* $\mathbb{F}_q$, that is, the group of invertible 2×2 matrices with coefficients in $\mathbb{F}_q$ and consider the following subgroups:

$$B = \left\{ \begin{pmatrix} a & b \\ 0 & d \end{pmatrix} : a, d \in \mathbb{F}_q^*, b \in \mathbb{F}_q \right\} \quad \text{(the \textit{Borel} subgroup)}$$

$$C = \left\{ \begin{pmatrix} a & \eta b \\ b & a \end{pmatrix} : a, b \in \mathbb{F}_q, (a, b) \neq (0, 0) \right\} \quad \text{(the \textit{Cartan} subgroup)}$$

$$D = \left\{ \begin{pmatrix} a & 0 \\ 0 & d \end{pmatrix} : a, d \in \mathbb{F}_q^* \right\} \quad \text{(the subgroup of \textit{diagonal} matrices)}$$

$$U = \left\{ \begin{pmatrix} 1 & b \\ 0 & 1 \end{pmatrix} : b \in \mathbb{F}_q \right\} \quad \text{(the subgroup of } \textit{unipotent} \text{ matrices)}$$

$$Z = \left\{ \begin{pmatrix} a & 0 \\ 0 & a \end{pmatrix} : a \in \mathbb{F}_q^* \right\} \quad \text{(the } \textit{center}\text{)}$$

where $\mathbb{F}_q^*$ denotes the multiplicative subgroup of $\mathbb{F}_q$ consisting of all non-zero elements, and η is a generator of the multiplicative group $\mathbb{F}_q^*$. As for subgroup C, we suppose that q is odd (for q even we refer to [17, Section 6.8]): then we have the isomorphism

$$\begin{aligned} C &\longrightarrow \mathbb{F}_{q^2}^* \\ \begin{pmatrix} a & \eta b \\ b & a \end{pmatrix} &\mapsto a + ib, \end{aligned} \tag{3.31}$$

where $\mathbb{F}_{q^2}$ is the quadratic extension of $\mathbb{F}_q$ and $\pm i$ are the square roots (in $\mathbb{F}_{q^2}$) of η.

An irreducible $\mathrm{GL}(2, \mathbb{F}_q)$-representation (ρ, V) such that the subspace V^U of U-invariant vectors is trivial is called a *cuspidal representation*.

Theorem 3.9 *Let χ be a non-trivial character of the (Abelian) group $K = U$. Then $\mathrm{Ind}_K^G \chi$ is multiplicity-free.*

In order to prove the above theorem we shall make use of the so-called *Bruhat decomposition* of G:

$$G = B \bigsqcup UwB \tag{3.32}$$

where $w = \begin{pmatrix} 0 & 1 \\ 1 & 0 \end{pmatrix}$ (this follows from elementary calculations, cf. [17, Lemma 14.2.4.(iv)]).

Proof of Theorem 3.9 We first observe that U is a normal subgroup of B and that one has $B = \bigsqcup_{d \in D} dU = \bigsqcup_{d \in D} UdU$. From (3.32) we then get

$$G = \left(\bigsqcup_{d \in D} dU \right) \bigsqcup \left(\bigsqcup_{d \in D} UwdU \right) = \left(\bigsqcup_{d \in D} UdU \right) \bigsqcup \left(\bigsqcup_{d \in D} UwdU \right).$$

As a consequence, we can take $\mathscr{S} := D \bigsqcup wD$ as a complete set of representatives for the double K-cosets in G. Moreover, it is easy to check that $dUd^{-1} \cap U = U$ and that $wdUd^{-1}w \cap U = \{1_G\}$ for all $d \in D$. As a consequence, cf. Theorem 3.8(2), we have that $\mathscr{S}_0 = Z \bigsqcup wD = \mathscr{S} \setminus (D \setminus Z)$. From Theorem 3.8(3) we deduce that every function $f \in \mathscr{H}(G, K, \psi)$ vanishes on $\bigsqcup_{d \in D \setminus Z} dU$.

Consider the map $\tau \colon G \to G$ defined by setting

$$\tau\left(\begin{pmatrix} a & b \\ c & d \end{pmatrix}\right) = \begin{pmatrix} d & b \\ c & a \end{pmatrix}$$

for all $\begin{pmatrix} a & b \\ c & d \end{pmatrix} \in G$. It is easy to check that $\tau(g_1 g_2) = \tau(g_2)\tau(g_1)$ and $\tau^2(g) = g$ for all $g_1, g_2, g \in G$. Thus, τ is an involutive antiautomorphism of G.

Let $f \in \mathscr{H}(G, K, \psi)$. We claim that

$$f(\tau(g)) = f(g) \quad \text{for all } g \in G. \tag{3.33}$$

In order to show (3.33), we recall that f is supported in $\bigsqcup_{s \in Z \bigsqcup wD} UsU$ and observe that τ fixes all the elements of the subgroup U. As a consequence, it suffices to show that τ also fixes all elements in $Z \bigsqcup wD$. This is a simple calculation:

$$\tau(wd) = \tau\left(\begin{pmatrix} 0 & 1 \\ 1 & 0 \end{pmatrix}\begin{pmatrix} a & 0 \\ 0 & b \end{pmatrix}\right) = \tau\left(\begin{pmatrix} 0 & b \\ a & 0 \end{pmatrix}\right) = \begin{pmatrix} 0 & b \\ a & 0 \end{pmatrix} = wd$$

for all $d = \begin{pmatrix} a & 0 \\ 0 & b \end{pmatrix} \in D$. On the other hand, it is obvious that $\tau(z) = z$ for all $z \in Z$. This proves the claim. As a consequence, by virtue of Proposition 3.1, we have that the Hecke algebra $\mathscr{H}(G, K, \psi)$ is commutative. From Theorem 3.1 we deduce that the induced representation $\mathrm{Ind}_K^G \chi$ is multiplicity-free. □

Remark 3.4 The conditions in the Bump–Ginzburg criterion are trivially satisfied, because $\tau(k) = k$ for all $k \in K$ and $\tau(s) = s$ for all $s \in \mathscr{S}_0$.

3.6 A Frobenius–Schur Theorem for Multiplicity-Free Triples

Let G be a finite group. Recall that the *conjugate* of a G-representation (σ, W) is the G-representation (σ', W') where W' is the dual of W and $[\sigma'(g)w'](w) = w'[\sigma(g^{-1})w]$ for all $g \in G$, $w \in W$, and $w' \in W'$. The matrix coefficients of the conjugate representation are the conjugate of the matrix coefficients, in formulae:

$$u_{i,j}^{\sigma'} = \overline{u_{i,j}^{\sigma}} \tag{3.34}$$

for all $i, j = 1, 2, \ldots, d_\sigma$ (cf. [11, Equation (9.14)]).

One then says that σ is *self-conjugate* provided $\sigma \sim \sigma'$; this is in turn equivalent to the associated character χ^σ being real-valued. When σ is not self-conjugate, one says that it is *complex*. The class of self-conjugate G-representations splits into two subclasses according to the associated matrix coefficients of the representation σ being real-valued or not: in the first case, one says that σ is *real*, in the second case σ is termed *quaternionic*.

A fundamental theorem of Frobenius and Schur provides a criterion for determining the type of a given irreducible G-representation σ, namely

$$\frac{1}{|G|}\sum_{g\in G}\chi^{\sigma}(g^2) = \begin{cases} 1 & \text{if } \sigma \text{ is real} \\ -1 & \text{if } \sigma \text{ is quaternionic} \\ 0 & \text{if } \sigma \text{ is complex} \end{cases} \tag{3.35}$$

see, for instance, [11, Theorem 9.7.7].

Let now $K \leq G$ be a subgroup, (θ, V) be an irreducible K-representation, and suppose that (G, K, θ) is a multiplicity-free triple. In the following we prove a generalization of the Frobenius–Schur theorem for spherical representations.

Theorem 3.10 *Let (G, K, θ) be a multiplicity-free triple and suppose that (σ, W) is a spherical representation (i.e., $\sigma \in \widehat{G}$ and $\sigma \preceq \mathrm{Ind}_K^G\theta$). Then we have*

$$\frac{d_\sigma}{|G|}\sum_{g\in G}\phi^{\sigma}(g^2) = \begin{cases} 1 & \textit{if } \sigma \textit{ is real} \\ -1 & \textit{if } \sigma \textit{ is quaternionic} \\ 0 & \textit{if } \sigma \textit{ is complex.} \end{cases} \tag{3.36}$$

Proof We first fix an orthonormal basis $\{u_i : i = 1, 2, \ldots, d_\sigma\}$ for W as in the paragraph preceding Theorem 3.7 so that $\phi^\sigma = u^\sigma_{1,1}$. We then have

$$\begin{aligned} \frac{1}{|G|}\sum_{g\in G}\phi^{\sigma}(g^2) &= \frac{1}{|G|}\sum_{g\in G}\overline{u^\sigma_{1,1}(g^2)} \\ (\text{by } (1.16)) &= \frac{1}{|G|}\sum_{g\in G}\sum_{h=1}^{d_\sigma}\overline{u^\sigma_{1,h}(g)u^\sigma_{h,1}(g)}. \end{aligned} \tag{3.37}$$

If σ is real, then $\overline{u^\sigma_{1,h}(g)} = u^\sigma_{1,h}(g)$ for all $h = 1, 2, \ldots, d_\sigma$ and $g \in G$, so that, by (1.14),

$$\frac{1}{|G|}\sum_{g\in G}\overline{u^\sigma_{1,h}(g)u^\sigma_{h,1}(g)} = \frac{1}{|G|}\langle u^\sigma_{1,h}, u^\sigma_{h,1}\rangle_{L(G)} = \frac{1}{d_\sigma}\delta_{1,h}$$

and (3.37) yields $\frac{d_\sigma}{|G|}\sum_{g\in G}\phi^\sigma(g^2) = 1$.

If σ is complex then, by (3.34),

$$\frac{1}{|G|}\sum_{g\in G}\overline{u^{\sigma}_{1,h}(g)u^{\sigma}_{h,1}(g)} = \frac{1}{|G|}\langle u^{\sigma'}_{1,h}, u^{\sigma}_{h,1}\rangle_{L(G)} = 0,$$

since $\sigma \not\sim \sigma'$ and (1.14) applies. Thus in this case (3.37) yields $\frac{d_\sigma}{|G|}\sum_{g\in G}\phi^{\sigma}(g^2) = 0$.

Suppose, finally, that σ is quaternionic. Then, see [11, Lemma 9.7.6], we can find a $d_\sigma \times d_\sigma$ complex matrix W such that $W\overline{W} = \overline{W}W = -I$ and $\overline{U(g)} = WU(g)W^*$, where $U(g) = \left(u^{\sigma}_{i,j}(g)\right)_{i,j=1}^{d_\sigma}$, for all $g \in G$. Then, for every $h = 1, 2, \ldots, d_\sigma$ we have

$$\begin{aligned}\frac{d_\sigma}{|G|}\sum_{g\in G}\overline{u^{\sigma}_{1,h}(g)u^{\sigma}_{h,1}(g)} &= \frac{d_\sigma}{|G|}\sum_{g\in G}\sum_{j,\ell=1}^{d_\sigma} w_{1,\ell}u^{\sigma}_{\ell,j}(g)\overline{w_{h,j}}\overline{u^{\sigma}_{h,1}(g)}\\ &= \sum_{j,\ell=1}^{d_\sigma} w_{1,\ell}\overline{w_{h,j}}\frac{d_\sigma}{|G|}\langle u^{\sigma}_{\ell,j}, u^{\sigma}_{h,1}\rangle_{L(G)}\\ \text{(by (1.14))} &= w_{1,h}\overline{w_{h,1}}\end{aligned}$$

and (3.37) yields

$$\frac{d_\sigma}{|G|}\sum_{g\in G}\phi^{\sigma}(g^2) = \sum_{h=1}^{d_\sigma} w_{1,h}\overline{w_{h,1}} = -1,$$

since $W\overline{W} = -I$. □

Remark 3.5 As mentioned at the end of Sect. 3.1, the twisted Frobenius–Schur theorem (cf. [15, Section 9]) deserves to be analyzed in the present setting.

Chapter 4
The Case of a Normal Subgroup

In this chapter we consider triples of the form (G, N, θ) in the particular case when the subgroup $N \leq G$ is normal.

It is straightforward to check that $\mathrm{Ind}_N^G \iota_N$, the induced representation of the trivial representation of N, is equivalent to the regular representation of the quotient group G/N. Therefore, if (G, N) is a Gelfand pair, its analysis is equivalent to the study of the representation theory of the quotient group G/N (which is necessarily Abelian).

In this chapter we treat the general case, namely, $\mathrm{Ind}_N^G \theta$, where $\theta \in \widehat{N}$ is arbitrary. Now, G acts by conjugation on $\widehat{N}$, and we denote by $I_G(\theta)$ the stabilizer of θ, called the *inertia group* (cf. Sect. 4.3). We then study the commutant $\mathrm{End}_G(\mathrm{Ind}_N^G \theta)$ of $\mathrm{Ind}_N^G \theta$.

From Clifford theory (cf. [12, Theorem 2.1(2)] and [14, Theorem 1.3.2(ii)]) it is known that $\dim \mathrm{End}_G(\mathrm{Ind}_N^G \theta) = |I_G(\theta)/N|$. In Theorem 4.1 we will show that, indeed, $\mathrm{End}_G(\mathrm{Ind}_N^G \theta)$ is isomorphic to the algebra $L(I_G(\theta)/N)$ equipped with a modified convolution product. Moreover, in Sect. 4.5 we shall study in detail this Hecke algebra and its associated spherical functions. Most of this chapter is indeed devoted to the general case, where we do not assume that $\mathrm{Ind}_N^G \theta$ is multiplicity-free. This is quite natural and has some interest on its own. In the last section we finally examine the multiplicity-free case and we prove that if $\mathrm{End}_G(\mathrm{Ind}_N^G \theta)$ is commutative, then $I_G(\theta)/N$ is Abelian and $\mathrm{End}_G(\mathrm{Ind}_N^G \theta) \cong L(I_G(\theta)/N)$ (where the latter is the usual group algebra); therefore, also for general representations induced from normal subgroups, the multiplicity-free case reduces to the analysis on an Abelian group. For more precise statements, see Theorems 4.3 and 4.4.

T. Ceccherini-Silberstein et al., *Gelfand Triples and Their Hecke Algebras*,
Lecture Notes in Mathematics 2267, https://doi.org/10.1007/978-3-030-51607-9_4

4.1 Unitary Cocycles

Let H be a finite group. Recall that the torus $\mathbb{T}$ is the quotient group $\mathbb{R}/\mathbb{Z}$.

Definition 4.1 A *unitary cocycle* on H is a map $\tau\colon H \times H \to \mathbb{T}$ such that

$$\tau(1_H, h) = \tau(h, 1_H) = 1_{\mathbb{T}} \qquad \text{(normalization)} \tag{4.1}$$

and

$$\tau(h_1h_2, h_3)\tau(h_1, h_2) = \tau(h_1, h_2h_3)\tau(h_2, h_3) \qquad \text{(cocycle identity)} \tag{4.2}$$

for all $h, h_1, h_2, h_3 \in H$.

We denote by $\mathscr{C}(\mathbb{T}, H)$ the set of all unitary cocycles.

It is easy to prove that $\mathscr{C}(\mathbb{T}, H)$ is an Abelian group under pointwise multiplication. Moreover, if a function $\rho\colon H \to \mathbb{T}$ satisfies the condition $\rho(1_H) = 1$, then $\tau_\rho\colon H \times H \to \mathbb{T}$, defined by setting

$$\tau_\rho(h_1, h_2) = \rho(h_1h_2)\left[\rho(h_1)\rho(h_2)\right]^{-1}, \tag{4.3}$$

for all $h_1, h_2 \in H$, is a unitary cocycle, called a *coboundary*. We denote by $\mathscr{B}(\mathbb{T}, H)$ the set of all coboundaries: it is a subgroup of $\mathscr{C}(\mathbb{T}, H)$. The corresponding quotient group $\mathscr{H}^2(\mathbb{T}, H) = \mathscr{C}(\mathbb{T}, H)/\mathscr{B}(\mathbb{T}, H)$ is called the *second cohomology group of H with values in* $\mathbb{T}$. The elements of $\mathscr{H}^2(\mathbb{T}, H)$ are called *cocycle classes* and two cocycles belonging to the same class are said to be *cohomologous*.

Note that if $\tau \in \mathscr{C}(\mathbb{T}, H)$, then, setting $h_1 = h_3 = h$ and $h_2 = h^{-1}$ in (4.2) and using (4.1), we get that $\tau(h, h^{-1}) = \tau(h^{-1}, h)$, for all $h \in H$. In particular, a unitary cocycle is said to be *equalized* (cf. [48]) if $\tau(h, h^{-1}) = 1$ for all $h \in H$. Every unitary cocycle τ is cohomologous to an equalized cocycle τ'. Indeed, if $\rho\colon H \to \mathbb{T}$ is defined by setting $\rho(1_H) = 1$ and $\rho(h) = \rho(h^{-1}) = \alpha$, where $\alpha \in \mathbb{T}$ satisfies $\alpha^2 = \tau(h, h^{-1})^{-1}$, for all $h \in H$, then one immediately checks that the function $\tau'\colon H \times H \to \mathbb{T}$, defined by setting

$$\tau'(h_1, h_2) = \rho(h_1)\rho(h_2)\rho(h_1h_2)^{-1}\tau(h_1, h_2), \tag{4.4}$$

for all $h_1, h_2 \in H$, is an equalized unitary cocycle cohomologous to τ.

Lemma 4.1 *If $\tau \in \mathscr{C}(\mathbb{T}, H)$ is equalized, then for all $g, h \in H$ we have:*

$$\tau(g, h)^{-1} = \tau(g^{-1}, gh) = \tau(h^{-1}, g^{-1}). \tag{4.5}$$

Moreover, for all $k, r, s \in H$ we have

$$\tau(k^{-1}s, s^{-1}r)\tau(k^{-1}, s) = \tau(k^{-1}, r)\tau(r^{-1}, s), \tag{4.6}$$

Proof Let $g, h, k, r, s \in H$. Setting $h_1 = g^{-1}$, $h_2 = g$ and $h_3 = h$ in (4.2), we get

$$\tau(g^{-1}, gh)\tau(g, h) = \tau(1_H, h)\tau(g^{-1}, g) = 1,$$

and this proves the first equality in (4.5). Similarly,

$$\begin{aligned} \tau(g,h)\tau(h^{-1},g^{-1}) &= \tau(g,h)\tau(gh,h^{-1}g^{-1})\tau(h^{-1},g^{-1}) \\ (h_1 = g, h_2 = h, \text{ and } h_3 = h^{-1}g^{-1} \text{ in (4.2)}) \quad &= \tau(g,g^{-1})\tau(h,h^{-1}g^{-1})\tau(h^{-1},g^{-1}) \\ (h_1 = h, h_2 = h^{-1}, \text{ and } h_3 = g^{-1} \text{ in (4.2)}) \quad &= \tau(1,g^{-1})\tau(h,h^{-1}) \\ &= 1, \end{aligned}$$

yields the second equality in (4.5).

Finally, we have

$$\tau(k^{-1}s, s^{-1}r)\tau(k^{-1}, s) = \tau(k^{-1}, r)\tau(s, s^{-1}r) = \tau(k^{-1}, r)\tau(r^{-1}, s),$$

where the first equality may be obtained by setting $h_1 = k^{-1}, h_2 = s$, and $h_3 = s^{-1}r$ in (4.2), and the second equality follows from the second equality in (4.5), by setting $g = s^{-1}$ and $h = r$. □

4.2 Cocycle Convolution

In this section we introduce a convolution product on $L(H)$ modified by means of an equalized unitary cocycle in $\mathscr{C}(\mathbb{T}, H)$ (see also [5, 44] where, for a similar algebra, the authors use the term *twisted convolution*).

Let $\eta \in \mathscr{C}(\mathbb{T}, H)$ be an equalized unitary cocycle. Given $f_1, f_2 \in L(H)$ we define their *η-cocycle convolution* by setting

$$[f_1 *_\eta f_2](k) = \sum_{h \in H} f_1(h^{-1}k) f_2(h)\eta(k^{-1}, h),$$

for all $k \in H$. Also, as usual, we let $f \mapsto f^*$ denote the involution on $L(H)$ defined by setting $f^*(h) = \overline{f(h^{-1})}$ for all $h \in H$.

Proposition 4.1 *The space $L(H)$ with the η-*cocycle convolution *and the involution defined above is an involutive, unital, associative algebra, which we denote by $L(H)_\eta$.*

Proof We first prove that the η-cocycle convolution is associative.
Let $f_1, f_2, f_3 \in L(H)$ and $k \in H$. Then we have

$$\begin{aligned}
\left[\left(f_1 *_\eta f_2\right) *_\eta f_3\right](k) &= \sum_{s\in H}\sum_{h\in H} f_1(h^{-1}s^{-1}k)f_2(h)f_3(s)\eta(k^{-1}s, h)\eta(k^{-1}, s)\\
(h = s^{-1}r) \qquad &= \sum_{s\in H}\sum_{r\in H} f_1(r^{-1}k)f_2(s^{-1}r)f_3(s)\eta(k^{-1}s, s^{-1}r)\eta(k^{-1}, s)\\
(\text{by } (4.6)) \qquad &= \sum_{s\in H}\sum_{r\in H} f_1(r^{-1}k)f_2(s^{-1}r)f_3(s)\eta(k^{-1}, r)\eta(r^{-1}, s)\\
&= \left[f_1 *_\eta \left(f_2 *_\eta f_3\right)\right](k).
\end{aligned}$$

This proves associativity of the convolution product $*_\eta$. Moreover,

$$\begin{aligned}
[f_1^* *_\eta f_2^*](k) &= \sum_{h\in H} \overline{f_1(k^{-1}h)f_2(h^{-1})\eta(k^{-1}, h)}\\
(\text{setting } s = k^{-1}h \text{ and by } (4.5)) \qquad &= \sum_{s\in H} \overline{f_1(s)}\,\overline{f_2(s^{-1}k^{-1})}\,\overline{\eta(k, s)}\\
&= [f_2 *_\eta f_1]^*(k).
\end{aligned}$$

This shows that $L(H)_\eta$ is involutive.
We leave it to the reader to check that the identity element is δ_{1_H}. □

4.3 The Inertia Group and Unitary Cocycle Representations

We recall some basic facts on Clifford theory; we refer to [12, 14] for more details and further results.

Let G be a finite group and suppose that $N \trianglelefteq G$ is a normal subgroup. Also let (θ, V) be an irreducible N-representation. Given $g \in G$, we denote by $({}^g\theta, V)$ the N-representation defined by setting

$$ {}^g\theta(n) = \theta(g^{-1}ng) \tag{4.7}$$

for all $n \in N$ (cf. (2.8)). This is called the *g-conjugate* representation of θ.

Observe that (4.7) defines a left action of G on $\widehat{N}$, i.e., ${}^{1_G}\theta = \theta$ and ${}^{g_1g_2}\theta = {}^{g_1}({}^{g_2}\theta)$ for all $g_1, g_2 \in G$. The stabilizer of this action is the subgroup

$$ I_G(\theta) = \{h \in G : {}^h\theta \sim \theta\}, \tag{4.8}$$

which is called the *inertia group* of θ. Note that $N \trianglelefteq I_G(\theta)$. Finally, we fix $Q \subseteq N$ a complete set of representatives for the cosets of N in $I_G(\theta)$ such that $1_G \in Q$, so that we have

$$I_G(\theta) = \bigsqcup_{q \in Q} qN = \bigsqcup_{q \in Q} Nq. \tag{4.9}$$

From now on, our exposition is based on [32, Section XII.1]; see also [18]. The same material is treated in [21, 43, 44] (where unitarity is not assumed) under the name of projective representations. Actually, we only need some specific portions of the theory but expressed in our language, so that we include complete proofs.

Since ${}^q\theta \sim \theta$ for all $q \in Q$, there exists a unitary operator $\Theta(q) \in \mathrm{End}(V)$ such that

$${}^q\theta(n) = \theta(q^{-1}nq) = \Theta(q)^{-1}\theta(n)\Theta(q), \tag{4.10}$$

for all $n \in N$. By Schur's lemma $\Theta(q)$ is determined up to a unitary multiplicative constant; we fix an arbitrary such a choice but we assume that $\Theta(1_G) = I_V$ (see also Remark 4.1). Then, on the whole of $I_G(\theta)$ (cf. (4.9)) we set

$$\Theta(nq) = \theta(n)\Theta(q) \tag{4.11}$$

$n \in N$ and $q \in Q$. Therefore, for all $h = nq \in I_G(\theta)$ and $m \in N$, we have:

$$\begin{aligned}\theta(h^{-1}mh) &= \theta(q^{-1}n^{-1}mnq) = \Theta(q)^{-1}\theta(n^{-1})\theta(m)\theta(n)\Theta(q) \\ &= \Theta(h)^{-1}\theta(m)\Theta(h),\end{aligned} \tag{4.12}$$

and

$$\Theta(mh) = \Theta(mnq) = \theta(mn)\Theta(q) = \theta(m)\theta(n)\Theta(q) = \theta(m)\Theta(h). \tag{4.13}$$

It follows that, for all $h, k \in I_G(\theta)$ and $n \in N$,

$$\Theta(h)^{-1}\Theta(k)^{-1}\theta(n)\Theta(k)\Theta(h) = \theta(h^{-1}k^{-1}nkh) = \Theta(kh)^{-1}\theta(n)\Theta(kh),$$

that is, $\Theta(kh)\Theta(h)^{-1}\Theta(k)^{-1}\theta(n) = \theta(n)\Theta(kh)\Theta(h)^{-1}\Theta(k)^{-1}$, so that, by Schur's lemma, there exists a constant $\tau(k, h) \in \mathbb{T}$ such that:

$$\Theta(kh) = \tau(k, h)\Theta(k)\Theta(h). \tag{4.14}$$

A map $I_G(\theta) \ni h \mapsto \Theta(h)$ (where each $\Theta(h)$ is unitary and τ is a unitary cocycle) satisfying (4.14) is called a *unitary τ-representation* (or a *cocycle* representation). We shall prove that τ is a unitary cocycle in Proposition 4.2. The notions of invariant subspaces, (unitary) equivalence, and irreducibility may be easily introduced for cocycle representations and we leave the corresponding details to the reader.

Lemma 4.2 *The maps* $\tau\colon I_G(\theta)\times I_G(\theta)\to\mathbb{T}$ *and* $\Theta\colon I_G(\theta)\to\mathrm{End}(V)$ *satisfy the following identities:*

$$\Theta(k)^* \equiv \Theta(k)^{-1} = \tau(k,k^{-1})\Theta(k^{-1}) \tag{4.15}$$

and

$$\Theta(h)\theta(m)\Theta(h)^{-1} = \theta(hmh^{-1}), \tag{4.16}$$

for all $k,h\in I_G(\theta), m\in N$. *Moreover, for* $k=nq_1, h=mq_2\in I_G(\theta), q_1,q_2\in Q, n,m\in N$,

$$\tau(k,h)I_V = \Theta(q_2)^{-1}\Theta(q_1)^{-1}\Theta(q_1q_2). \tag{4.17}$$

Proof Since $\Theta(1_G) = I_V$, from (4.14) with $h=k^{-1}$ we deduce that $I_V = \tau(k,k^{-1})\Theta(k)\Theta(k^{-1})$, and the first identity follows. We now prove the second identity:

$$\begin{aligned}\Theta(h)\theta(m)\Theta(h)^{-1} &= \Theta(h^{-1})^{-1}\theta(m)\Theta(h^{-1}) && \text{(by (4.15))}\\ &= \theta(hmh^{-1}) && \text{(by (4.12))}.\end{aligned}$$

Finally,

$$\begin{aligned}\Theta(kh) &= \Theta(nq_1mq_2) = \Theta(nq_1mq_1^{-1}\cdot q_1q_2)\\ \text{(by (4.11) and (4.16))} \quad &= \theta(n)\Theta(q_1)\theta(m)\Theta(q_1)^{-1}\Theta(q_1q_2)\end{aligned} \tag{4.18}$$

so that

$$\begin{aligned}\tau(k,h)I_V &= \Theta(h)^{-1}\Theta(k)^{-1}\cdot\Theta(kh) && \text{(by (4.14))}\\ &= \Theta(q_2)^{-1}\theta(m)^{-1}\Theta(q_1)^{-1}\theta(n)^{-1}\cdot && \text{(by (4.11))}\\ &\quad\cdot\theta(n)\Theta(q_1)\theta(m)\Theta(q_1)^{-1}\Theta(q_1q_2) && \text{(by (4.18))}\\ &= \Theta(q_2)^{-1}\Theta(q_1)^{-1}\Theta(q_1q_2).\end{aligned}$$

□

In Remark 4.1 we will give a simplified version of (4.15).

Proposition 4.2

(1) *The function* τ *defined by* (4.14) *is a 2-cocycle on* $I_G(\theta)$.
(2) *The cocycle* τ *is bi-*N*-invariant:* $\tau(nhn',mkm')=\tau(h,k)$, *for all* $h,k\in I_G(\theta)$ *and* $n,n',m,m'\in N$.

Proof

(1) From the condition $\Theta(1_G) = I_V$ and (4.14) it follows that $\tau(k, 1_G) = \tau(1_G, h) = 1$, for $k, h \in I_G(\theta)$. Moreover, from (4.14) it also follows that, for all $h_1, h_2, h_3 \in I_G(\theta)$,

$$\begin{aligned}\tau(h_1h_2, h_3)\tau(h_1, h_2)I_V &= \Theta(h_1h_2h_3)\Theta(h_3)^{-1}\Theta(h_1h_2)^{-1}\\ &\quad\cdot\Theta(h_1h_2)\Theta(h_2)^{-1}\Theta(h_1)^{-1}\\ &= \Theta(h_1h_2h_3)\Theta(h_2h_3)^{-1}\tau(h_2, h_3)\Theta(h_1)^{-1}\\ &= \tau(h_1, h_2h_3)\tau(h_2, h_3)I_V.\end{aligned}$$

(2) Let $h, k \in I_G(\theta)$. Given $n, m \in N$, from

$$\begin{aligned}\tau(nh, k)\Theta(nh)\Theta(k) &= \Theta(nhk) && \text{(by (4.14))}\\ &= \tau(h, k)\theta(n)\Theta(h)\Theta(k) && \text{(by (4.13) and (4.14))}\\ &= \tau(h, k)\Theta(nh)\Theta(k) && \text{(by (4.13))}\end{aligned}$$

it follows that $\tau(nh, k) = \tau(h, k)$, while from

$$\begin{aligned}\tau(h, k)\Theta(hmk) &= \tau(h, k)\tau(h, mk)\Theta(h)\Theta(mk) && \text{(by (4.14))}\\ &= \tau(h, k)\tau(h, mk)\Theta(h)\theta(m)\Theta(k) && \text{(by (4.13))}\\ &= \tau(h, k)\tau(h, mk)\theta(hmh^{-1})\Theta(h)\Theta(k) && \text{(by (4.16))}\\ &= \tau(h, mk)\theta(hmh^{-1})\Theta(hk) && \text{(by (4.14))}\\ &= \tau(h, mk)\Theta(hmk), && \text{(by (4.13))}\end{aligned}$$

we deduce that $\tau(h, mk) = \tau(h, k)$. This shows left-N-invariance of τ. Given $n', m' \in N$ we can find $n, m \in N$ such that $hn' = nh$ and $km' = mk$. Using left-N-invariance we have

$$\tau(hn', km') = \tau(nh, mk) = \tau(h, k)$$

and right-N-invariance follows as well. □

Corollary 4.1 *For all $h \in I_G(\theta)$ and $n \in N$*

$$\Theta(hn) = \Theta(h)\theta(n). \tag{4.19}$$

Proof We have $\Theta(hn) = \tau(h, n)\Theta(h)\theta(n)$ and $\tau(h, n) = \tau(h, 1_G) = 1$. □

As a consequence of Proposition 4.2.(2), we define a unitary cocycle $\eta \in \mathscr{C}(\mathbb{T}, I_G(\theta)/N)$ by setting

$$\eta(hN, kN) = \tau(h, k), \tag{4.20}$$

for all $h, k \in I_G(\theta)$.

For the following proposition we keep the notation above, referring to a fixed $Q \subset I_G(\theta)$ complete set of representatives of the N-cosets in $I_G(\theta)$. We want to show that η is independent of the choices of Q and Θ.

Proposition 4.3 *Let $Q' \subset I_G(\theta)$ be another complete set of representatives of the N-cosets in $I_G(\theta)$ (possibly, $Q' = Q$), and denote by $\Theta'(q')$, $q' \in Q'$, a family of unitary operators on V satisfying*

$$^{q'}\theta(n) = \Theta'(q')^{-1}\theta(n)\Theta'(q') \tag{4.21}$$

for all $n \in N$ (cf. (4.10)). Also set

$$\Theta'(nq') = \theta(n)\Theta'(q') \tag{4.22}$$

for all $n \in N$ and $q' \in Q'$ (cf. (4.11)). Then the corresponding unitary cocycle η' is cohomologous to η.

Proof For each $q' \in Q'$ there exist unique $n_{q'} \in N$ and $q \in Q$ such that

$$q' = n_{q'}q. \tag{4.23}$$

By (4.11) we then have

$$\Theta(q') = \Theta(n_{q'}q) = \theta(n_{q'})\Theta(q). \tag{4.24}$$

Moreover,

$$\begin{aligned} {}^{q'}\theta(n) &= \theta(q^{-1}n_{q'}^{-1}nn_{q'}q) \\ &= \Theta(q)^{-1}\theta(n_{q'}^{-1}nn_{q'})\Theta(q) \\ &= \Theta(q)^{-1}\theta(n_{q'})^{-1}\theta(n)\theta(n_{q'})\Theta(q) \\ \text{(by (4.24))} \quad &= \Theta(q')^{-1}\theta(n)\Theta(q'). \end{aligned} \tag{4.25}$$

Comparing (4.21) and (4.25), from Schur's lemma we deduce that there exists $\psi \colon I_G(\theta) \to \mathbb{T}$ such that

$$\Theta'(h) = \psi(h)\Theta(h)$$

for all $h \in I_G(\theta)$. Note that ψ is bi-N-invariant: $\psi(nq') = \psi(q') = \psi(q'n)$ for all $n \in N$ and $q' \in Q'$, where the second equality follows from Corollary 4.1. In particular, $\psi(1_G) = 1$. Then for $k, h \in I_G(\theta)$

$$\begin{aligned}\tau'(k,h)I_V &= \Theta'(h)^{-1}\Theta'(k)^{-1}\Theta'(kh) \\ &= \Theta(h)^{-1}\Theta(k)^{-1}\Theta(kh)\psi(h)^{-1}\psi(k)^{-1}\psi(kh) \\ &= \tau(k,h)\psi(h)^{-1}\psi(k)^{-1}\psi(kh)I_V,\end{aligned} \tag{4.26}$$

that is, τ and τ' are cohomologous. Finally, setting $\rho(hN) = \psi(h)$, for all $h \in N$ (cf. Eq. (4.20)) we deduce that η and η' are cohomologous: (4.26) becomes $\eta'(kN, hN) = \eta(kN, hN)\rho(hN)^{-1}\rho(kN)^{-1}\rho(khN)$. □

Remark 4.1 Let $\Theta\colon I_G(\theta) \to \mathrm{End}(V)$ be as in (4.10) and denote by $\tau\colon I_G(\theta) \times I_G(\theta) \to \mathbb{T}$ the corresponding unitary cocycle as in (4.14). For each $k \in I_G(\theta)$ let $\tau'(k) = \tau'(k^{-1})$ be a square root of $\tau(k, k^{-1}) = \tau(k^{-1}, k)$ and set

$$\Theta'(k) = \tau'(k)\Theta(k).$$

Then, from (4.15) we deduce

$$\Theta'(k)^{-1} = \left(\tau'(k)\Theta(k)\right)^{-1} = \tau'(k)^{-1}\tau(k,k^{-1})\Theta(k^{-1}) = \tau'(k^{-1})\Theta(k^{-1}) = \Theta'(k^{-1}).$$

Note that Θ and Θ' give rise to cohomologous cocycles η and η', by Proposition 4.3, but η' is equalized; see also (4.4).

4.4 A Description of the Hecke Algebra $\tilde{\mathscr{H}}(G, N, \theta)$

In the present setting, the Hecke algebra in Sect. 2.1 is made up of all functions $F\colon G \to \mathrm{End}(V)$ such that

$$F(ngm) = \theta(m^{-1})F(g)\theta(n^{-1}), \qquad \text{for all} \quad g \in G, n, m \in N. \tag{4.27}$$

We also suppose that the unitary cocycle η is equalized; see Remark 4.1.

In the following theorem we prove the normal subgroup version of Mackey's formula for invariants.

Theorem 4.1 *For each $F \in \tilde{\mathscr{H}}(G, N, \theta)$ there exists $f \in L(I_G(\theta)/N)$ such that:*

$$F(h) = \frac{1}{|N|}\Theta(h)^* f(hN), \tag{4.28}$$

for all $h \in I_G(\theta)$, while $F(g) = 0$ for $g \notin I_G(\theta)$. Moreover, the map

$$\begin{array}{rcl} \Phi : L(I_G(\theta)/N)_\eta & \longrightarrow & \widetilde{\mathscr{H}}(G, N, \theta) \\ f & \longmapsto & F, \end{array} \tag{4.29}$$

where F is as in (4.28), is a $*$-isomorphism of algebras such that $\sqrt{|N|}\Phi$ is isometric.

Proof From (4.13) and (4.19) it follows that the function $I_G(\theta) \ni h \mapsto \Theta(h)^* \in \mathrm{End}(V)$ belongs to $\widetilde{\mathscr{H}}(G, N, \theta)$. Moreover, from (4.12) it follows that for each $h \in I_G(\theta)$ the operator $\Theta(h)^*$ belongs to $\mathrm{Hom}_N({}^{h^{-1}}\theta, \theta)$ and in fact spans it: this follows from the fact that this space is one-dimensional (by Schur's lemma). Recall also that by (4.15) and by Remark 4.1 we may suppose that

$$\Theta(h)^* \equiv \Theta(h)^{-1} = \Theta(h^{-1}) \tag{4.30}$$

for all $h \in I_G(\theta)$. Similarly, from (4.27) it follows that if $F \in \widetilde{\mathscr{H}}(G, N, \theta)$ then

$$\theta(n)F(g) = F(gn^{-1}) = F(gn^{-1}g^{-1} \cdot g) = F(g)\left[{}^{g^{-1}}\theta\right](n),$$

that is, $F(g) \in \mathrm{Hom}_N({}^{g^{-1}}\theta, \theta)$. In particular, $F(g) = 0$ if $g \notin I_G(\theta)$ and there exists $f \in L(I_G(\theta)/N)$ such that F is of the form (4.28). Indeed, $F(h) = \widetilde{f}(h)\Theta(h)^*$ for some constant $\widetilde{f}(h) \in \mathbb{C}$ and from

$$\theta(n^{-1})\widetilde{f}(h)\Theta(h)^* = \theta(n^{-1})F(h) = F(hn) = \widetilde{f}(hn)\Theta(hn)^* = \theta(n^{-1})\widetilde{f}(hn)\Theta(h)^*$$

we get that $\widetilde{f}$ in N-invariant; we then set $f(hN) = |N|\widetilde{f}(h)$ (where $|N|$ is a normalization constant). Conversely, any function of the form (4.28) clearly belongs to $\widetilde{\mathscr{H}}(G, N, \theta)$.

Let $f_1, f_2 \in L(I_G(\theta)/N)$ and $k \in I_G(\theta)$. Let us show that the map Φ preserves the convolution. Taking into account (2.2) and (4.14)), we have:

$$\begin{aligned}
[\Phi(f_1) * \Phi(f_2)](k) &= \frac{1}{|N|^2} \sum_{h \in I_G(\theta)} f_1(h^{-1}kN) f_2(hN)\Theta(k^{-1}h)\Theta(h^{-1}) \\
&= \frac{1}{|N|^2} \sum_{h \in I_G(\theta)} f_1(h^{-1}kN) f_2(hN)\Theta(k^{-1}) \\
&\quad \cdot \tau(k^{-1}, h)\Theta(h)\Theta(h^{-1}) \\
&=_* \frac{1}{|N|}\Theta(k)^* \sum_{hN \in I_G(\theta)/N} f_1(h^{-1}N \cdot kN) f_2(hN)\eta(k^{-1}N, hN) \\
&= \frac{1}{|N|}\Theta(k)^*\left[f_1 *_\eta f_2\right](kN) \\
&= \Phi(f_1 *_\eta f_2)(k),
\end{aligned}$$

where $=_*$ follows from (4.30) and (4.20). Similarly,

$$\begin{aligned}\langle \Phi(f_1), \Phi(f_2)\rangle_{\widetilde{\mathscr{H}}(G,N,\theta)} &= \sum_{h\in I_G(\theta)} \frac{1}{\dim V}\mathrm{tr}\left[\Phi(f_2)(h)^*\Phi(f_1)(h)\right] \text{ (by (2.3))}\\ &= \sum_{h\in I_G(\theta)} \frac{1}{|N|^2\dim V} f_1(hN)\overline{f_2(hN)}\mathrm{tr}\left[\Theta(h)\Theta(h)^*\right]\\ &= \sum_{hN\in I_G(\theta)/N} \frac{1}{|N|} f_1(hN)\overline{f_2(hN)}\\ &= \frac{1}{|N|}\langle f_1, f_2\rangle_{L(I_G(\theta)/N)},\end{aligned}$$

so that the map $\sqrt{|N|}\Phi$ is an isometry.

We are only left to show that Φ preserves the involutions. Let $f \in L(I_G(\theta)/N)$ and $k \in I_G(\theta)$. Keeping in mind (2.4) and (4.30) we have

$$\begin{aligned}\Phi(f)^*(k) = [\Phi(f)(k^{-1})]^* &= \frac{1}{|N|}\Theta(k^{-1})\overline{f(k^{-1}N)}\\ &= \frac{1}{|N|}\Theta(k)^* f^*(kN) = \Phi(f^*)(k).\end{aligned}$$

□

Corollary 4.2 $\dim \mathrm{End}_G(\mathrm{Ind}_N^G\theta) = |I_G(\theta)/N|$.

Moreover, from Theorem 2.1, we can also deduce the following

Corollary 4.3 *The composition $\xi \circ \Phi \colon L(I_G(\theta)/N)_\eta \to \mathrm{End}_G(\mathrm{Ind}_N^G V)$ is indeed an isometric $*$-isomorphism of algebras.*

4.5 The Hecke Algebra $\widetilde{\mathscr{H}}(G, N, \psi)$

In order to describe, in the present framework, the Hecke algebra $\mathscr{H}(G, N, \psi)$ (cf. Sect. 2.2), we introduce the function $\Psi \colon I_G(\theta) \to \mathbb{C}$ by setting

$$\Psi(h) = \langle v, \Theta(h)v\rangle \tag{4.31}$$

for all $h \in I_G(\theta)$, where $v \in V$ is a fixed vector with $\|v\| = 1$, and $\Theta \colon I_G(\theta) \to \mathrm{End}(V)$ is as in Sect. 4.3. Clearly, in the notation of (2.24), we have, keeping in mind (4.11),

$$\psi(n) = \frac{d_\theta}{|N|}\Psi(n) \tag{4.32}$$

for all $n \in N$.

Theorem 4.2 *We have*

$$\Psi = \Psi * \psi = \psi * \Psi = \psi * \Psi * \psi \tag{4.33}$$

and, for all $k, h \in I_G(\theta)$,

$$\sum_{n \in N} \Psi(knh)\overline{\psi(n)} = \overline{\tau(k,h)}\Psi(k)\Psi(h). \tag{4.34}$$

Proof Let $k \in I_G(\theta)$. Then

$$\begin{aligned}
[\Psi * \psi](k) &= \sum_{n \in N} \Psi(kn^{-1})\psi(n) \\
&= \sum_{n \in N} \langle v, \Theta(kn^{-1})v\rangle \frac{d_\theta}{|N|} \cdot \langle v, \theta(n)v\rangle \\
\text{(by (4.19) and (4.30))} \quad &= \left\langle \Theta(k^{-1})v, \frac{d_\theta}{|N|} \sum_{n \in N} \langle \theta(n)v, v\rangle \theta(n^{-1})v \right\rangle \\
\text{(by (2.23))} \quad &= \langle \Theta(k^{-1})v, v\rangle \\
&= \Psi(k).
\end{aligned}$$

This shows the first equality in (4.33), the other ones follow after similar computations.

In order to show (4.34), let $k, h \in I_G(\theta)$ and $n \in N$. We first note that

$$\begin{aligned}
\Theta(knh) &= \Theta(knk^{-1} \cdot kh) \\
\text{(by (4.13))} \quad &= \theta(knk^{-1})\Theta(kh) \\
\text{(by (4.16))} \quad &= \Theta(k)\theta(n)\Theta(k)^{-1}\Theta(kh) \\
\text{(by (4.14))} \quad &= \tau(k,h)\Theta(k)\theta(n)\Theta(h).
\end{aligned} \tag{4.35}$$

We deduce that

$$\begin{aligned}
\sum_{n \in N} \Psi(knh)\overline{\psi(n)} &= \sum_{n \in N} \langle v, \Theta(knh)v\rangle \frac{d_\theta}{|N|} \langle \theta(n)v, v\rangle \\
\text{(by (4.35))} \quad &= \overline{\tau(k,h)} \left\langle \Theta(k^{-1})v, \frac{d_\theta}{|N|} \sum_{n \in N} \langle \theta(n^{-1})v, v\rangle \theta(n)\Theta(h)v \right\rangle \\
\text{(by (2.22))} \quad &= \overline{\tau(k,h)} \left\langle \Theta(k^{-1})v, \langle \Theta(h)v, v\rangle v \right\rangle \\
&= \overline{\tau(k,h)}\Psi(k)\Psi(h).
\end{aligned}$$

□

For every $\phi \in \widehat{I_G(\theta)/N}$ denote by $\widetilde{\phi}$ its *inflation* to $I_G(\theta)$. This is defined by setting $\widetilde{\phi}(h) = \phi(hN)$ for all $h \in I_G(\theta)$ (i.e. we compose the projection map $I_G(\theta) \to I_G(\theta)/N$ with $\phi\colon I_G(\theta)/N \to \mathrm{End}(V_\phi)$).

Proposition 4.4 *We have*

$$\mathscr{H}(G, N, \psi) = \{\widetilde{\phi}\Psi : \phi \in L(I_G(\theta)/N)\}$$

where

$$[\widetilde{\phi}\Psi](h) = \phi(hN)\Psi(h)$$

for all $h \in H$. In particular, every $f \in \mathscr{H}(G, N, \psi)$ vanishes outside $I_G(\theta)$.

Proof From Theorem 2.2.(3) we deduce that $\mathscr{H}(G, N, \psi)$ is made up of all $S_v F$, where F is as in (4.28) with f replaced by ϕ, that is, for all $h \in I_G(\theta)$,

$$\begin{aligned}
[S_v F](h) &= d_\theta \langle F(h)v, v\rangle \\
&= \frac{d_\theta}{|N|}\phi(hN)\langle v, \Theta(h)v\rangle \\
&= \frac{d_\theta}{|N|}\widetilde{\phi}(h)\Psi(h).
\end{aligned}$$

For the last statement, recall that a function $F \in \widetilde{\mathscr{H}}(G, N, \psi)$ vanishes outside $I_G(\theta)$, by Theorem 4.1. □

4.6 The Multiplicity-Free Case and the Spherical Functions

In this section we study the multiplicity-free case of a triple (G, N, θ) and determine the associated spherical functions.

We first introduce another notation from Clifford theory (see [12] and [14]):

$$\widehat{I}(\theta) = \{\xi \in \widehat{I_G(\theta)} : \xi \preceq \mathrm{Ind}_N^{I_G(\theta)}\theta\}.$$

Remark 4.2 Note that if (G, N, θ) is a multiplicity-free triple, then also $(I_G(\theta), N, \theta)$ is multiplicity-free. Indeed the commutant of $\mathrm{Ind}_N^G\theta$ coincides, by virtue of Theorems 2.1 and 4.1, with $L(I_G(\theta)/N)_\eta$. Since $I_{I_G(\theta)}(\theta) = I_G(\theta)$, the previous argument shows that the latter is also the commutant of $\mathrm{Ind}_N^{I_G(\theta)}\theta$. In other words, the commutant depends only on $I_G(\theta)$. Alternatively, this also follows from transitivity of induction.

Theorem 4.3 $\mathrm{Ind}_N^G\theta$ *decomposes without multiplicities if and only if the following conditions are satisfied:*

(i) $I_G(\theta)/N$ *is Abelian;*
(ii) $\mathrm{Res}_N^{I_G(\theta)}\xi = \theta$ *for all* $\xi \in \widehat{I}(\theta)$.

Proof If $\mathrm{Ind}_N^G\theta$ is multiplicity-free then also $\mathrm{Ind}_N^{I_G(\theta)}\theta$ is multiplicity-free (see Remark 4.2). Then

$$\mathrm{Ind}_N^{I_G(\theta)}\theta \sim \bigoplus_{\xi\in\widehat{I}(\theta)} \xi. \tag{4.36}$$

By Frobenius reciprocity

$$\mathrm{Hom}_{I_G(\theta)}(\theta, \mathrm{Res}_N^{I_G(\theta)}\mathrm{Ind}_N^{I_G(\theta)}\theta) \cong \mathrm{End}_{I_G(\theta)}(\mathrm{Ind}_N^{I_G(\theta)}\theta).$$

By taking dimensions, the multiplicity of θ in $\mathrm{Res}^{I_G(\theta)}\mathrm{Ind}_N^{I_G(\theta)}\theta$ equals

$$\dim \mathrm{End}_{I_G(\theta)}(\mathrm{Ind}_N^{I_G(\theta)}\theta) = |I_G(\theta)/N| \tag{4.37}$$

by Corollary 4.2. Again by computing dimension, since $\dim \mathrm{Ind}_N^{I_G(\theta)}\theta = |I_G(\theta)/N| \cdot \dim\theta$, we deduce that indeed

$$\mathrm{Res}_N^{I_G(\theta)}\mathrm{Ind}_N^{I_G(\theta)}\theta \sim |I_G(\theta)/N|\theta. \tag{4.38}$$

But (4.36) and (4.37) force $|\widehat{I}(\theta)| = |I_G(\theta)/N|$ (cf. (1.27) and (1.28) for $m_\sigma = 1$) and therefore, by (4.38), necessarily $\mathrm{Res}_N^{I_G(\theta)}\xi = \theta$ for all $\xi \in \widehat{I}(\theta)$, and (ii) is proved.

Then, for all $\xi \in \widehat{I}(\theta)$, we have

$$\begin{aligned}\mathrm{Ind}_N^{I_G(\theta)}\theta &\sim \mathrm{Ind}_N^{I_G(\theta)}(\theta \otimes \iota_N)\\ \text{(by (ii))} &\sim \mathrm{Ind}_N^{I_G(\theta)}[(\mathrm{Res}_N^{I_G(\theta)}\xi) \otimes \iota_N]\\ \text{(by [17, Theorem 11.1.16])} &\sim \xi \otimes \mathrm{Ind}_N^G\iota_N\\ &\sim \xi \otimes \widetilde{\lambda}.\end{aligned}$$

From the decomposition of the regular representation of $I_G(\theta)/N$

$$\lambda \sim \bigoplus_{\phi\in\widehat{I_G(\theta)/N}} d_\phi\phi$$

we deduce that

$$\mathrm{Ind}_N^{I_G(\theta)}\theta \sim \bigoplus_{\phi \in \widehat{I_G(\theta)/N}} d_\phi \xi \otimes \widetilde{\phi} \tag{4.39}$$

so that (4.36) forces $d_\phi \equiv 1$ for all $\phi \in \widehat{I_G(\theta)/N}$ and therefore $I_G(\theta)/N$ is Abelian (see [17, Exercise 10.3.16]). So that also (i) is proved.

We now assume (i) and (ii). Then in (4.10) we can take

$$\Theta = \xi \tag{4.40}$$

choosing any $\xi \in \widehat{I}(\theta)$. Therefore in (4.14) we have $\tau \equiv 1$ so that also $\eta \equiv 1$ in Theorem 4.1. Then (cf. Corollary 4.3)

$$\mathrm{End}_{I_G(\theta)}(\mathrm{Ind}_N^{I_G(\theta)}\theta) \cong L(I_G(\theta)/N)$$

with the usual convolution structure, so that it is commutative and $\mathrm{Ind}_N^{I_G(\theta)}\theta$ is multiplicity-free (cf. Proposition 1.3) and therefore (see Remark 4.2) also $\mathrm{Ind}_N^G\theta$ is multiplicity-free. □

Remark 4.3 The proof of Theorem 4.3 is based on Theorem 4.1 and its corollaries, that is, on the normal subgroup version of Mackey's formula for invariants. Another proof, based on Clifford Theory, which is a normal subgroup version of Mackey's lemma, is in [18]. In that paper we plan to develop a Fourier analysis of the algebra $L(H)_\eta$ (see Sect. 4.2) in the noncommutative case and therefore also of $\mathrm{End}_G(\mathrm{Ind}_N^G\theta)$ when $\mathrm{Ind}_N^G\theta$ is not multiplicity-free (cf. [61] for a different approach to the analysis of $\mathrm{Ind}_K^G\theta$, where $K \leq G$ is not necessarily a normal subgroup and the representation decomposes with multiplicities).

In what follows, $\chi \in \widehat{I_G(\theta)/N}$ denotes a character of the Abelian group $I_G(\theta)/N$ and $\widetilde{\chi}$ its inflation to $I_G(\theta)$, that is, $\widetilde{\chi}(h) = \chi(hN)$ for all $h \in I_G(\theta)$, while $\overline{\chi}$ is the *complex conjugate*, that is, $\overline{\chi}(hN) = \overline{\chi(hN)}$ (as a complex number).

Recall (cf. Definition 3.2) that $\mathscr{S}(G, N, \psi) \subseteq \mathscr{H}(G, N, \psi)$ denotes the set of all spherical functions associated with the multiplicity-free triple (G, N, θ).

Proposition 4.5 *Let (G, N, θ) be a multiplicity-free triple. Then*

$$\mathscr{S}(G, N, \psi) = \{\widetilde{\chi}\Psi : \chi \in \widehat{I_G(\theta)/N}\}.$$

Proof Recall that a function $\phi \in L(G)$ is spherical if and only if it satisfies (3.12) and, moreover, if this is the case, ϕ must be of the form $\widetilde{\varphi}\Psi$ for some $\varphi \in L(I_G(\theta)/N)$, by Proposition 4.4.

We then show that, given $\varphi\colon I_G(\theta)/N \to \mathbb{T}$, then $\phi = \widetilde{\varphi}\Psi$ satisfies (3.12) if and only if φ is a character. This follows immediately after comparing

$$\sum_{n\in N} \phi(knh)\overline{\psi(n)} = \sum_{n\in N} \varphi(knhN)\Psi(knh)\overline{\psi(n)} = \varphi(khN)\Psi(k)\Psi(h)$$

(where the last equality follows from (4.34) with $\tau \equiv 1$) and the last expression is equal to $\varphi(kN)\Psi(k)\varphi(hN)\Psi(h)$ if and only if φ is a character belonging to $\widehat{I_G(\theta)/N}$. □

Theorem 4.4 *Suppose that (G, N, θ) is a multiplicity-free triple. Fix $\xi \in \widehat{I}(\theta)$ and $v \in V$ with $\|v\| = 1$, and set $\Psi(h) = \langle v, \xi(h)v\rangle$ for all $h \in I_G(\theta)$ (cf. (4.31) and (4.40)). Then the following hold.*

(1) *The map*

$$\begin{array}{ccc} L(I_G(\theta)/N) & \to & \mathscr{H}(G, N, \psi) \\ f & \mapsto & \frac{d_\theta}{|I_G(\theta)|}\widetilde{f}\Psi \end{array}$$

is an isomorphism of commutative algebras.

(2)

$$\mathrm{Ind}_N^G\theta \sim \bigoplus_{\chi\in\widehat{I_G(\theta)/N}} \mathrm{Ind}_{I_G(\theta)}^G(\xi \otimes \widetilde{\chi}) \tag{4.41}$$

is the decomposition of $\mathrm{Ind}_N^G\theta$ into irreducibles (that is, into the spherical representations of the multiplicity-free triple (G, N, θ)).

(3) *The spherical function associated with $\mathrm{Ind}_{I_G(\theta)}^G(\xi \otimes \widetilde{\chi})$ is given by*

$$\phi^\chi = \widetilde{\chi}\Psi.$$

Proof

(1) This follows from Proposition 4.4, by taking into account that now $\eta \equiv 1$ (see the proof of Theorem 4.3), that $I_G(\theta)/N$ is commutative, and that $\Psi * \Psi = \frac{|I_G(\theta)|}{d_\theta}\Psi$ (cf. (1.15)). Note also that, in order to compute the convolution in $\mathscr{H}(G, N, \psi)$, we may use the expression for the spherical functions in Proposition 4.5, (1.15), and (4.40) applied to the $I_G(\theta)$-representations $\xi \otimes \chi$, with $\chi \in \widehat{I_G(\theta)/N}$. Alternatively, this isomorphism is the composition of the map (4.29) with the map $S_v\colon \mathscr{H}(G, K, \theta) \longrightarrow L(G)$ in Theorem 2.2.(3) (cf. the proof of Proposition 4.5).

(2) The decomposition (4.41) follows from transitivity of induction and (4.39), taking into account that $d_\phi = 1$ and using χ to denote a generic character of $I_G(\theta)/N$. Moreover,

$$\dim \mathrm{End}_G(\mathrm{Ind}_N^G \theta) = |I_G(\theta)/N| = \dim \mathrm{End}_{I_G(\theta)}(\mathrm{Ind}_N^{I_G(\theta)} \theta)$$

(see Corollary 4.2, Corollary 4.3, and Remark 4.2) forces each representation in the right hand side of (4.41) be irreducible.

(3) From Proposition 4.5 applied to $(I_G(\theta), N, \theta)$ and (4.31) (with Θ replaced by ξ; see (4.40)) we deduce that the function

$$H \ni h \mapsto \overline{\widetilde{\chi}}(h)\Psi(h) = \overline{\widetilde{\chi}}(h)\langle v, \xi(h)v\rangle = \langle v, \widetilde{\chi}(h)\xi(h)v\rangle$$

is the spherical functions of $(I_G(\theta), N, \theta)$ associated with $\xi \otimes \widetilde{\chi}$ (cf. (3.24); now L_σ is trivial because $\mathrm{Res}_N^{I_G(\theta)} \xi \sim \theta$ act on the same space V_θ). From Proposition 4.5 it follows that $\mathscr{S}(G, N, \psi) \equiv \mathscr{S}(I_G(\theta), N, \psi)$ (recall also that each element of $\mathscr{H}(G, N, \psi)$ vanishes outside $I_G(\theta)$; cf. Proposition 4.4). Moreover, from the characterization of the spherical representations in Theorem 3.6 and the characterization (or definition) of induced representations in [17, Proposition 11.1.2] it follows that the spherical function associated with $\mathrm{Ind}_{I_G(\theta)}^G(\xi \otimes \widetilde{\chi})$ coincides with that associated with $\xi \otimes \widetilde{\chi}$. $\square$

Chapter 5
Harmonic Analysis of the Multiplicity-Free Triple $(\mathrm{GL}(2, \mathbb{F}_q), C, \nu)$

In this chapter we study a family of multiplicity-free triples on $\mathrm{GL}(2, \mathbb{F}_q)$ that generalize the well known Gelfand pair associated with the finite hyperbolic plane (see [69, Chapters 19, 20, 21, and 23]). We suppose that q is an odd prime power (cf. Sect. 3.5) and we denote by $\widehat{\mathbb{F}_q^*}$ (respectively $\widehat{\mathbb{F}_{q^2}^*}$) the multiplicative characters of $\mathbb{F}_q$ (respectively $\mathbb{F}_{q^2}$). If $\nu \in \widehat{\mathbb{F}_{q^2}^*}$, we may think of ν as a one-dimensional representation of the Abelian group C by setting (cf. (3.31))

$$\nu \begin{pmatrix} \alpha & \eta\beta \\ \beta & \alpha \end{pmatrix} = \nu(\alpha + i\beta),$$

where $\alpha, \beta \in \mathbb{F}_q$, $(\alpha, \beta) \neq (0, 0)$, η is a generator of the multiplicative group $\mathbb{F}_q^*$, and $\pm i$ are the square roots (in $\mathbb{F}_{q^2}$) of η.

5.1 The Multiplicity-Free Triple $(\mathrm{GL}(2, \mathbb{F}_q), C, \nu)$

We denote by $\mathrm{Aff}(\mathbb{F}_q)$ the (*general*) *affine group* (*of degree one*) over $\mathbb{F}_q$, that is, the subgroup of $\mathrm{GL}(2, \mathbb{F}_q)$ defined by

$$\mathrm{Aff}(\mathbb{F}_q) = \left\{ \begin{pmatrix} a & b \\ 0 & 1 \end{pmatrix} : a \in \mathbb{F}_q^*,\ b \in \mathbb{F}_q \right\}.$$

Note that $\mathrm{Aff}(\mathbb{F}_q)$ acts on $\mathbb{F}_q \equiv \left\{ \begin{pmatrix} x \\ 1 \end{pmatrix} : x \in \mathbb{F}_q \right\}$ by multiplication:

$$\begin{pmatrix} a & b \\ 0 & 1 \end{pmatrix} \begin{pmatrix} x \\ 1 \end{pmatrix} = \begin{pmatrix} ax + b \\ 1 \end{pmatrix}.$$

T. Ceccherini-Silberstein et al., *Gelfand Triples and Their Hecke Algebras*, Lecture Notes in Mathematics 2267, https://doi.org/10.1007/978-3-030-51607-9_5

The following is an elementary, but quite useful lemma.

Lemma 5.1

(1) *Every* $\begin{pmatrix} \alpha & \beta \\ \gamma & \delta \end{pmatrix} \in \mathrm{GL}(2, \mathbb{F}_q)$ *may be written uniquely as the product of a matrix in* $\mathrm{Aff}(\mathbb{F}_q)$ *and a matrix in* C, *namely*

$$\begin{pmatrix} \alpha & \beta \\ \gamma & \delta \end{pmatrix} = \begin{pmatrix} x & y \\ 0 & 1 \end{pmatrix} \begin{pmatrix} a & \eta b \\ b & a \end{pmatrix},$$

where $a = \delta$, $b = \gamma$, $x = \frac{\alpha\delta - \beta\gamma}{\delta^2 - \eta\gamma^2}$, $y = \frac{\beta\delta - \alpha\gamma\eta}{\delta^2 - \eta\gamma^2}$.

(2) *Every* $\begin{pmatrix} a & b \\ c & d \end{pmatrix} \in \mathrm{GL}(2, \mathbb{F}_q)$ *may be written uniquely as the product of a matrix in* C *by a matrix in* $\mathrm{Aff}(\mathbb{F}_q)$, *namely*

$$\begin{pmatrix} a & b \\ c & d \end{pmatrix} = \begin{pmatrix} \alpha & \eta\beta \\ \beta & \alpha \end{pmatrix} \begin{pmatrix} x & y \\ 0 & 1 \end{pmatrix}.$$

(3) *For all* $x, y, \alpha, \beta \in \mathbb{F}_q$ *with* $x \neq 0$ *and* $(\alpha, \beta) \neq (0, 0)$ *we have*

$$\begin{pmatrix} x & y \\ 0 & 1 \end{pmatrix} \begin{pmatrix} \alpha & \eta\beta \\ \beta & \alpha \end{pmatrix} = \begin{pmatrix} u & \eta v \\ v & u \end{pmatrix} \begin{pmatrix} a & b \\ 0 & 1 \end{pmatrix}, \tag{5.1}$$

where

$$\begin{aligned} u &= x(x\alpha + y\beta)\frac{\alpha^2 - \beta^2\eta}{(x\alpha + y\beta)^2 - \eta\beta^2} \\ v &= x\beta\frac{\alpha^2 - \eta\beta^2}{(x\alpha + y\beta)^2 - \eta\beta^2} \\ a &= \frac{(x\alpha + y\beta)^2 - \eta\beta^2}{x(\alpha^2 - \eta\beta^2)} \\ b &= \frac{(x\alpha + y\beta)(y\alpha + \eta\beta x) - \eta\alpha\beta}{x(\alpha^2 - \eta\beta^2)}. \end{aligned} \tag{5.2}$$

Proof The proof consists just of elementary but tedious computations. Alternatively, the reader may quickly check the formulas (5.2) with computer algebra, e.g., with the mathematics software Maple.

(1) Since

$$\begin{pmatrix} x & y \\ 0 & 1 \end{pmatrix} \begin{pmatrix} a & \eta b \\ b & a \end{pmatrix} = \begin{pmatrix} xa + yb & \eta bx + ya \\ b & a \end{pmatrix},$$

we immediately get $b = \gamma$ and $a = \delta$. Moreover, the linear system

$$\begin{cases} x\delta + y\gamma & = \alpha \\ x\eta\gamma + y\delta & = \beta \end{cases}$$

may be solved by Cramer's rule, yielding the expression for x and y in the statement (just note that η is not a square, because q is odd, and therefore $\delta^2 - \eta\gamma^2 \neq 0$).

(2) This follows from (1), simply by taking inverses.

(3) We just sketch the calculations. (5.1) is equivalent to the system

$$\begin{cases} ua & = x\alpha + y\beta \\ ub + v\eta & = x\beta\eta + y\alpha \\ va & = \beta \\ vb + u & = \alpha. \end{cases}$$

For the moment, assume $\beta \neq 0$. Then from the third and the fourth equation, we find

$$a = v^{-1}\beta \quad b = v^{-1}\alpha - v^{-1}u$$

and, from the first equation, we deduce that

$$u = (x\alpha\beta^{-1} + y)v.$$

A substitution of these expressions in the second equation leads to the explicit formula for v. Then one can derive the expressions for u and a, and finally the expression for b. The final formulas also include the case $\beta = 0$. □

Proposition 5.1 *$(\mathrm{GL}(2, \mathbb{F}_q), C, \nu)$ is a multiplicity-free triple for all $\nu \in \widehat{\mathbb{F}^*_{q^2}}$.*

Proof We use the Bump–Ginzburg criterion (Theorem 3.3) with $\tau\colon \mathrm{GL}(2, \mathbb{F}_q) \to \mathrm{GL}(2, \mathbb{F}_q)$ given by

$$\tau\begin{pmatrix} a & b \\ c & d \end{pmatrix} = \begin{pmatrix} d & b \\ c & a \end{pmatrix}$$

for all $\begin{pmatrix} a & b \\ c & d \end{pmatrix} \in \mathrm{GL}(2, \mathbb{F}_q)$. It is easy to check that τ is an involutive antiautomorphism (cf. Theorem 3.9 and [17, Theorem 14.6.3]). Moreover, $\tau(k) = k$ for all $k \in C$ and, consequently, $\nu(\tau(k)) = \nu(k)$ for all $k \in C$ and $\nu \in \widehat{\mathbb{F}^*_{q^2}}$ (viewed, as remarked above, as a one-dimensional representation of C). It only remains to prove conditions (3.7) and (3.8). Actually, $\mathcal{S}$ is quite difficult to describe and we refer to [69, pp. 311–322], where a complete geometric description is developed. From our point of view, however, it is enough to know that $\mathscr{S}$ is a subset of $\mathrm{Aff}(\mathbb{F}_q)$: this follows from Lemma 5.1(1) and (2). Therefore we prove conditions (3.7) and (3.8) for all $s \in \mathrm{Aff}(\mathbb{F}_q)$. We shall see that, given $s = \begin{pmatrix} x & y \\ 0 & 1 \end{pmatrix}$, there exists $k = k(x, y) = \begin{pmatrix} \alpha & \eta\beta \\ \beta & \alpha \end{pmatrix}$ such that $\tau(s) = ksk^{-1}$, that is, (3.7) with $k_1 = k$ and $k_2 = k^{-1}$ (then (3.8) is trivially satisfied: in the present setting it becomes $\nu(k)\nu(k^{-1}) = \nu(1) = 1$). First of all, we note that

$$\tau(s) = \tau \begin{pmatrix} x & y \\ 0 & 1 \end{pmatrix} = \begin{pmatrix} 1 & y \\ 0 & x \end{pmatrix}$$

so that $ks = \tau(s)k$ becomes

$$\begin{pmatrix} \alpha & \eta\beta \\ \beta & \alpha \end{pmatrix} \begin{pmatrix} x & y \\ 0 & 1 \end{pmatrix} = \begin{pmatrix} 1 & y \\ 0 & x \end{pmatrix} \begin{pmatrix} \alpha & \eta\beta \\ \beta & \alpha \end{pmatrix}$$

which is equivalent to

$$\alpha x = \alpha + \beta y.$$

Thus, if $x \neq 1$ (respectively $x = 1$), $k(x, y)$ is obtained by choosing β arbitrarily and then setting $\alpha = (x-1)^{-1}\beta y$ (respectively by choosing α arbitrarily and setting $\beta = 0$). □

Another proof of Proposition 5.1 is given in Remark 5.3.

Remark 5.1 The above result may be expressed by saying, in the terminology of [13, Section 2.1.2], that C is a *multiplicity-free* subgroup of $\mathrm{GL}(2, \mathbb{F}_q)$. This is equivalent (cf. [13, Theorem 2.1.10]) to $(G \times C, \widetilde{C})$ being a Gelfand pair, where $\widetilde{C} = \{(h, h) : H \in C\}$. We plan to study the Gelfand pair $(G \times C, \widetilde{C})$ in full details in a future paper. Other references on this and similar related constructions include [11, 59, Section 9.8], and [16].

5.2 Representation Theory of $\mathrm{GL}(2, \mathbb{F}_q)$: Parabolic Representations

We now recall some basic facts on the representation theory of the group $\mathrm{GL}(2, \mathbb{F}_q)$. We refer to [17, Chapter 14] for complete proofs. For simplicity, we set $G = \mathrm{GL}(2, \mathbb{F}_q)$.

The $(q-1)$ one-dimensional representations are of the form

$$\widehat{\chi}^0_\psi(g) = \psi(\det g)$$

for all $g \in G$, where $\psi \in \widehat{\mathbb{F}^*_q}$.

For $\psi_1, \psi_2 \in \widehat{\mathbb{F}^*_q}$ consider the one-dimensional representation of the Borel subgroup B given by

$$\chi_{\psi_1,\psi_2}\begin{pmatrix}\alpha & \beta\\ 0 & \delta\end{pmatrix} = \psi_1(\alpha)\psi_2(\delta) \tag{5.3}$$

for all $\begin{pmatrix}\alpha & \beta\\ 0 & \delta\end{pmatrix} \in B$. Then we set

$$\widehat{\chi}_{\psi_1,\psi_2} = \mathrm{Ind}_B^G \chi_{\psi_1,\psi_2}.$$

If $\psi_1 \neq \psi_2$ then $\widehat{\chi}_{\psi_1,\psi_2}$ is irreducible and $\widehat{\chi}_{\psi_1,\psi_2} \sim \widehat{\chi}_{\psi_3,\psi_4}$ if and only if $\{\psi_1, \psi_2\} = \{\psi_3, \psi_4\}$. If $\psi_1 = \psi_2 = \psi$ then

$$\widehat{\chi}_{\psi,\psi} = \widehat{\chi}^0_\psi \oplus \widehat{\chi}^1_\psi$$

where $\widehat{\chi}^0_\psi$ is the one-dimensional representation defined above and $\widehat{\chi}^1_\psi$ is a q-dimensional irreducible representation. The representations $\widehat{\chi}_{\psi_1,\psi_2}$ and $\widehat{\chi}^1_\psi$ are called *parabolic representations*: these are $(q-1)(q-2)/2$ representations of dimension $q+1$ and $q-1$ representations of dimension q, respectively.

5.3 Representation Theory of GL(2, $\mathbb{F}_q$): Cuspidal Representations

We now introduce the last class of irreducible representations of G. We need further preliminary results from [17, Chapter 7]. Let $\nu \in \widehat{\mathbb{F}^*_{q^2}}$ and set $\nu^\sharp = \mathrm{Res}^{\mathbb{F}^*_{q^2}}_{\mathbb{F}^*_q} \nu$ and, for $\alpha + i\beta \in \mathbb{F}^*_{q^2}$ (cf. (3.31)), set $\overline{\alpha + i\beta} = \alpha - i\beta$ (conjugation). Then, the map $\alpha + i\beta \mapsto \alpha - i\beta$ is precisely the unique non-trivial authomorphism of $\mathbb{F}_{q^2}$ that fixes each element of $\mathbb{F}_q$. Clearly, $\nu \mapsto \nu^\sharp$ is a group homomorphism and each $\psi \in \widehat{\mathbb{F}^*_q}$ is the image of $\frac{|\widehat{\mathbb{F}^*_{q^2}}|}{|\widehat{\mathbb{F}^*_q}|} = q+1$ characters of $\mathbb{F}^*_{q^2}$. For $\psi \in \widehat{\mathbb{F}^*_q}$ we set

$$\Psi(w) = \psi(w\overline{w}) \text{ for all } w \in \mathbb{F}^*_{q^2}. \tag{5.4}$$

Then $\Psi \in \widehat{\mathbb{F}^*_{q^2}}$ and we say that Ψ is *decomposable*. If $\nu \in \widehat{\mathbb{F}^*_{q^2}}$ but cannot be written in this form, it is called *indecomposable*. For $\nu \in \widehat{\mathbb{F}^*_{q^2}}$ we set $\overline{\nu}(w) = \nu(\overline{w})$ and we have that $\overline{\nu} \in \widehat{\mathbb{F}^*_{q^2}}$. Then, ν is indecomposable if and only if $\nu \neq \overline{\nu}$ (warning: as usual, $\overline{\nu(z)}$ will indicate the conjugate of the complex number $\nu(z)$).

Suppose now that $\nu \in \widehat{\mathbb{F}^*_{q^2}}$ is indecomposable and $\chi \in \widehat{\mathbb{F}_q}$ is a nontrivial *additive character* of $\mathbb{F}_q$. Following the monograph by Piatetski-Shapiro [53] (but we refer to [17, Chapter 7]) we introduce the *generalized Kloosterman sum* $j = j_{\chi,\nu}\colon \mathbb{F}_q^* \to \mathbb{C}$ defined by setting

$$j(x) = \frac{1}{q} \sum_{\substack{w \in \mathbb{F}^*_{q^2}: \\ w\bar{w} = x}} \chi(w + \bar{w})\nu(w)$$

for all $x \in \mathbb{F}_q^*$.

We will use repeatedly the following identities (cf. [17, Proposition 7.3.4 and Corollary 7.36]):

$$\sum_{z \in \mathbb{F}_q^*} j(xz)j(yz)\nu(z^{-1}) = \delta_{x,y}\nu(-x) \tag{5.5}$$

$$\sum_{z \in \mathbb{F}_q^*} j(xz)j(yz)\nu(z^{-1})\chi(z) = -\chi(-x-y)\nu(-1)j(xy) \tag{5.6}$$

for all $x, y \in \mathbb{F}_q^*$, and

$$\sum_{\alpha \in \mathbb{F}_q^*} \nu(-\alpha)\chi(\alpha^{-1}(z+\overline{z}))j(\alpha^{-2}z\overline{z}) = \nu(z) + \nu(\overline{z}) \tag{5.7}$$

for all $z \in \mathbb{F}^*_{q^2}$. Actually, the last identity was obtained during the computation of the characters of the cuspidal representations; cf. the end of the proof of the character table in [17, Section 14.9] (where $\delta = z' + \overline{z}$ and $\beta = -z\overline{z}$). Finally (cf. [17, Proposition 7.3.3]),

$$\overline{j(x)} = j(x)\overline{\nu(-x)}. \tag{5.8}$$

We now describe the cuspidal representation associated with an indecomposable character $\nu \in \mathbb{F}^*_{q^2}$: the representation space is $L(\mathbb{F}_q^*)$ and the representation ρ_ν is defined by setting, for all $g \in G$, $f \in L(\mathbb{F}_q^*)$, and $y \in \mathbb{F}_q^*$,

$$[\rho_\nu(g)f](y) = \nu(\delta)\chi(\delta^{-1}\beta y^{-1})f(\delta\alpha^{-1}y) \tag{5.9}$$

if $g = \begin{pmatrix} \alpha & \beta \\ 0 & \delta \end{pmatrix} \in B$, and

$$[\rho_\nu(g)f](y) = -\sum_{x\in\mathbb{F}_q^*} \nu(-\gamma x)\chi(\alpha\gamma^{-1}y^{-1}+\gamma^{-1}\delta x^{-1})j(\gamma^{-2}y^{-1}x^{-1}\det(g))f(x) \tag{5.10}$$

if $g = \begin{pmatrix} \alpha & \beta \\ \gamma & \delta \end{pmatrix} \in G \setminus B \equiv BwB$ (that is, if $\gamma \neq 0$).

Actually, $\rho_\nu \sim \rho_\mu$ if and only if $\nu = \mu$ or $\nu = \overline{\mu}$, so that we have $q(q-1)/2$ cuspidal representations.

Remark 5.2 Collecting from Sects. 5.2 and 5.3 the lists of all one-dimensional, parabolic, and cuspidal representations of $\mathrm{GL}(2, \mathbb{F}_q)$ together with their dimensions, we obtain that the sum of the squares of the dimensions of these irreducible representations, namely,

$$(q-1) + (q+1)^2 \cdot \frac{(q-1)(q-2)}{2} + q^2 \cdot (q-1) + (q-1)^2 \cdot \frac{q(q-1)}{2}$$

equals the order $(q^2-1)(q^2-q) = q(q+1)(q-1)^2$ of the group $\mathrm{GL}(2, \mathbb{F}_q)$ (cf. [17, Proposition 14.3.1]). This shows that there are no other irreducible representations of $\mathrm{GL}(2, \mathbb{F}_q)$.

5.4 The Decomposition of $\mathrm{Ind}_C^G \nu_0$

Theorem 5.1 *For all indecomposable characters $\nu_0 \in \widehat{\mathbb{F}_{q^2}^*}$ we have the decomposition*

$$\mathrm{Ind}_C^G \nu_0 \sim \left(\bigoplus_{\substack{\psi\in\widehat{\mathbb{F}_q^*}:\\ \psi^2=\nu_0^\sharp}} \widehat{\chi}_\psi^1 \right) \oplus \left(\bigoplus_{\substack{\{\psi_1,\psi_2\}\subseteq\widehat{\mathbb{F}_q^*}:\\ \psi_1\neq\psi_2,\\ \psi_1\psi_2=\nu_0^\sharp}} \widehat{\chi}_{\psi_1,\psi_2} \right) \oplus \left(\bigoplus_{\substack{\nu\in\widehat{\mathbb{F}_{q^2}^*} \text{ indecomposable}:\\ \nu^\sharp=\nu_0^\sharp\\ \nu\neq\nu_0,\overline{\nu_0}}} \rho_\nu \right). \tag{5.11}$$

Proof The fact that the above decomposition is multiplicity-free follows from Proposition 5.1. Actually, we use Frobenius reciprocity, so that we study the restrictions to C of all irreducible G-representations. From the proof of [17, Theorem 14.3.2], it follows that, for $\beta \neq 0$, the matrices

$$\begin{pmatrix} \alpha & \eta\beta \\ \beta & \alpha \end{pmatrix} \quad \text{and} \quad \begin{pmatrix} 0 & \eta\beta^2 - \alpha^2 \\ 1 & 2\alpha \end{pmatrix}$$

belong to the same conjugacy class of G. Moreover, if $z = \alpha + i\beta$ then $z + \overline{z} = 2\alpha$ and $-z\overline{z} = -\alpha^2 + \eta\beta^2$. From the character table in [17, Table 14.2, Section 14.9] we have, for $\psi, \psi_1, \psi_2 \in \widehat{\mathbb{F}_q^*}$ and $\nu \in \widehat{\mathbb{F}_{q^2}^*}$ indecomposable,

- $\widehat{\chi}_\psi^0 \begin{pmatrix} \alpha & \eta\beta \\ \beta & \alpha \end{pmatrix} = \begin{cases} \psi(z\overline{z}) & \text{if } \beta \neq 0 \\ \psi(\alpha^2) & \text{if } \beta = 0 \end{cases} \equiv \psi(z\overline{z});$
- $\chi^{\widehat{\chi}_\psi^1} \begin{pmatrix} \alpha & \eta\beta \\ \beta & \alpha \end{pmatrix} = \begin{cases} -\psi(z\overline{z}) & \text{if } \beta \neq 0 \\ q\psi(\alpha^2) & \text{if } \beta = 0; \end{cases}$
- $\chi^{\widehat{\chi}_{\psi_1,\psi_2}} \begin{pmatrix} \alpha & \eta\beta \\ \beta & \alpha \end{pmatrix} = \begin{cases} 0 & \text{if } \beta \neq 0 \\ (q+1)\psi_1(\alpha)\psi_2(\alpha) & \text{if } \beta = 0; \end{cases}$
- $\chi^{\rho_\nu} \begin{pmatrix} \alpha & \eta\beta \\ \beta & \alpha \end{pmatrix} = \begin{cases} -\nu(z) - \nu(\overline{z}) & \text{if } \beta \neq 0 \\ (q-1)\nu(\alpha) & \text{if } \beta = 0. \end{cases}$

Let now ν_0 be an indecomposable character of $\mathbb{F}_{q^2}^*$. We compute the multiplicity of ν_0 in the restriction $\mathrm{Res}_C^G\theta$, with $\theta \in \widehat{G}$, by means of the scalar product of characters of C (see [17, Proposition 10.2.18]). Note that $|C| = q^2 - 1$ and that the condition $\beta \neq 0$ is equivalent to $z \in \mathbb{F}_{q^2}^* \setminus \mathbb{F}_q^*$. The multiplicity of ν_0 in $\mathrm{Res}_C^G \widehat{\chi}_\psi^0$ is equal to

$$\frac{1}{q^2-1} \sum_{z \in \mathbb{F}_{q^2}^*} \psi(z\overline{z})\overline{\nu_0(z)} = \delta_{\Psi,\nu_0} = 0, \tag{5.12}$$

because ν_0 is indecomposable (Ψ is as in (5.4)).

The multiplicity of ν_0 in $\mathrm{Res}_C^G \widehat{\chi}_\psi^1$ is equal to

$$\begin{aligned}
&\frac{1}{q^2-1} \sum_{z \in \mathbb{F}_{q^2}^* \setminus \mathbb{F}_q^*} -\psi(z\overline{z})\overline{\nu_0(z)} + \frac{1}{q^2-1} \sum_{\alpha \in \mathbb{F}_q^*} q\psi(\alpha^2)\overline{\nu_0(\alpha)} \\
&= \frac{1}{q^2-1} \sum_{z \in \mathbb{F}_{q^2}^*} -\psi(z\overline{z})\overline{\nu_0(z)} + \frac{1}{q-1} \sum_{\alpha \in \mathbb{F}_q^*} \psi(\alpha^2)\overline{\nu_0(\alpha)} \\
&= -\delta_{\Psi,\nu_0} + \delta_{\psi^2,\nu_0^\sharp} \\
&= \delta_{\psi^2,\nu_0^\sharp},
\end{aligned} \tag{5.13}$$

where the last equality follows, once more, from the fact that ν_0 is indecomposable.

The multiplicity of ν_0 in $\mathrm{Res}^G_C \widehat{\chi}_{\psi_1,\psi_2}$ is equal to

$$\frac{1}{q^2-1}\sum_{\alpha\in\mathbb{F}_q^*}(q+1)\psi_1(\alpha)\psi_2(\alpha)\overline{\nu_0(\alpha)} = \delta_{\psi_1\psi_2,\nu_0^\sharp}. \tag{5.14}$$

Finally, the multiplicity of ν_0 in $\mathrm{Res}^G_C \rho_\nu$ is equal to

$$\begin{aligned}
&\frac{1}{q^2-1}\sum_{z\in\mathbb{F}_{q^2}^*\setminus\mathbb{F}_q^*}[-\nu(z)-\nu(\overline{z})]\overline{\nu_0(z)} + \frac{1}{q^2-1}\sum_{\alpha\in\mathbb{F}_q^*}(q-1)\nu(\alpha)\overline{\nu_0(\alpha)}\\
&= \frac{1}{q^2-1}\sum_{z\in\mathbb{F}_{q^2}^*}[-\nu(z)\overline{\nu_0(z)} - \overline{\nu}(z)\overline{\nu_0(z)}] + \frac{1}{q-1}\sum_{\alpha\in\mathbb{F}_q^*}\nu(\alpha)\overline{\nu_0(\alpha)}\\
&= -\delta_{\nu,\nu_0} - \delta_{\overline{\nu},\nu_0} + \delta_{\nu^\sharp,\nu_0^\sharp}\\
&= \begin{cases} 1 & \text{if } \nu^\sharp = \nu_0^\sharp \text{ and } \nu_0 \neq \nu, \overline{\nu}\\ 0 & \text{otherwise.}\end{cases}
\end{aligned} \tag{5.15}$$

As we alluded to at the beginning of the proof, the decomposition in (5.11) follows from (5.12), (5.13), (5.14), and (5.15) by invoking Frobenius reciprocity. □

Remark 5.3 Note that in the above proof we have, incidentally, recovered the fact that C is a multiplicity-free subgroup of G (the case of a decomposable $\Psi \in \widehat{\mathbb{F}_{q^2}^*}$ may be handled similarly). This second approach has the advantage of yielding also the decompositions into irreducible representations both of restriction and induced representations; however, it requires the knowledge of the representation theory of $\mathrm{GL}(2, \mathbb{F}_q)$.

5.5 Spherical Functions for $(\mathrm{GL}(2, \mathbb{F}_q), C, \nu_0)$: the Parabolic Case

In this section we derive an explicit formula for the spherical functions associated with the parabolic representations. We first find an explicit expression for the decomposition of the restriction to C of the parabolic representations. Note that the involved representations have already been determined in the proof of Theorem 5.1.

Theorem 5.2

(1) *Let $\psi_1, \psi_2 \in \widehat{\mathbb{F}_q^*}$, $\psi_1 \neq \psi_2$. For every $\nu \in \widehat{\mathbb{F}_{q^2}^*}$ such that $\nu^\sharp = \psi_1\psi_2$ define $F_\nu : G \to \mathbb{C}$ by setting*

$$F_\nu\left[\begin{pmatrix}\alpha & \eta\beta\\ \beta & \alpha\end{pmatrix}\begin{pmatrix}x & y\\ 0 & 1\end{pmatrix}\right] = \overline{\nu(\alpha+i\beta)\psi_1(x)} \tag{5.16}$$

(cf. Lemma 5.1(2)). Then F_ν belongs to $V_{\widehat{\chi}_{\psi_1,\psi_2}}$, the representation space of $\widehat{\chi}_{\psi_1,\psi_2}$. Moreover,

$$\lambda\begin{pmatrix}\alpha & \eta\beta\\ \beta & \alpha\end{pmatrix}F_\nu = \nu(\alpha+i\beta)F_\nu \tag{5.17}$$

for all $\alpha+i\beta\in\mathbb{F}_{q^2}^*$, and

$$V_{\widehat{\chi}_{\psi_1,\psi_2}} \cong \bigoplus_{\substack{\nu\in\mathbb{F}_{q^2}^*:\\ \nu^\sharp=\psi_1\psi_2}} \mathbb{C}F_\nu \tag{5.18}$$

is an explicit form of the decomposition

$$\mathrm{Res}^G_C\widehat{\chi}_{\psi_1,\psi_2} \sim \bigoplus_{\substack{\nu\in\mathbb{F}_{q^2}^*:\\ \nu^\sharp=\psi_1\psi_2}} \nu. \tag{5.19}$$

Finally, $\|F_\nu\|_{V_{\widehat{\chi}_{\psi_1,\psi_2}}} = \sqrt{q-1}$.

(2) *Let $\psi\in\widehat{\mathbb{F}_q^*}$ and, for every* indecomposable *$\nu\in\widehat{\mathbb{F}_{q^2}^*}$ such that $\nu^\sharp=\psi^2$, define $F_\nu\colon G\to\mathbb{C}$ by setting*

$$F_\nu\left[\begin{pmatrix}\alpha & \eta\beta\\ \beta & \alpha\end{pmatrix}\begin{pmatrix}x & y\\ 0 & 1\end{pmatrix}\right] = \overline{\nu(\alpha+i\beta)\psi(x)}.$$

Then F_ν belongs to $V_{\widehat{\chi}_\psi^1}$, the representation space of $\widehat{\chi}_\psi^1$,

$$\lambda\begin{pmatrix}\alpha & \eta\beta\\ \beta & \alpha\end{pmatrix}F_\nu = \nu(\alpha+i\beta)F_\nu$$

for all $\alpha+i\beta\in\mathbb{F}_{q^2}^*$. Moreover

$$V_{\widehat{\chi}_\psi^1} \cong \bigoplus_{\substack{\nu\in\mathbb{F}_{q^2}^*:\\ \nu\ \text{indecomposable}\\ \nu^\sharp=\psi^2}} \mathbb{C}F_\nu$$

is an explicit form of the decomposition

$$\mathrm{Res}^G_C\widehat{\chi}_\psi^1 \sim \bigoplus_{\substack{\nu\in\mathbb{F}_{q^2}^*:\\ \nu\ \text{indecomposable}\\ \nu^\sharp=\psi^2}} \nu.$$

Finally, $\|F_\nu\|_{V_{\widehat{\chi}_\psi^1}} = \sqrt{q-1}$.

Proof We just prove (1), because the proof of (2) is essentially the same.

We prove that F_ν belongs to the representation space of $\widehat{\chi}_{\psi_1,\psi_2} = \mathrm{Ind}_B^G \chi_{\psi_1,\psi_2}$ by verifying directly definition (1.19). For all $\begin{pmatrix} a & b \\ 0 & d \end{pmatrix} \in B$, we have

$$\begin{aligned}
F_\nu\left[\begin{pmatrix} \alpha & \eta b \\ \beta & \alpha \end{pmatrix}\begin{pmatrix} x & y \\ 0 & 1 \end{pmatrix}\begin{pmatrix} a & b \\ 0 & d \end{pmatrix}\right] &= F_\nu\left[\begin{pmatrix} \alpha & \eta\beta \\ \beta & \alpha \end{pmatrix}\begin{pmatrix} xa & xb+yd \\ 0 & d \end{pmatrix}\right] \\
&= F_\nu\left[\begin{pmatrix} \alpha d & \eta\beta d \\ \beta d & \alpha d \end{pmatrix}\begin{pmatrix} xad^{-1} & xbd^{-1}+y \\ 0 & 1 \end{pmatrix}\right] \\
\text{(by (5.16))} &= \overline{\nu(\alpha d + i\beta d)}\,\overline{\psi_1(xad^{-1})} \\
&= \overline{\nu(\alpha + i\beta)}\,\overline{\psi_1(x)}\cdot\overline{\nu(d)}\,\overline{\psi_1(ad^{-1})} \\
(\nu^\sharp = \psi_1\psi_2) &= F_\nu\left[\begin{pmatrix} \alpha & \eta\beta \\ \beta & \alpha \end{pmatrix}\begin{pmatrix} x & y \\ 0 & 1 \end{pmatrix}\right]\overline{\psi_1(a)\psi_2(d)}. \qquad (5.20)
\end{aligned}$$

Then (5.17) is obvious and, from it, (5.18) and (5.19) follow immediately. Finally, by (1.24),

$$\begin{aligned}
\|F_\nu\|^2_{V_{\widehat{\chi}_{\psi_1,\psi_2}}} &= \frac{1}{q(q-1)^2}\sum_{\alpha+i\beta\in\mathbb{F}_{q^2}^*}\sum_{x\in\mathbb{F}_q^*}\sum_{y\in\mathbb{F}_q}\overline{\nu(\alpha+i\beta)\psi_1(x)}\nu(\alpha+i\beta)\psi_1(x) \\
&= \frac{(q^2-1)q(q-1)}{q(q-1)^2} = q+1.
\end{aligned}$$

□

Remark 5.4 Recall that Mackey's lemma (cf. [13, Theorem 1.6.14], [17, Theorem 11.5.1], and [9]) states that if $H, K \leq G$ and (σ, V) is an irreducible K-representation, then $\mathrm{Res}_H^G \mathrm{Ind}_K^G V \cong \bigoplus_{s\in\mathscr{S}} \mathrm{Ind}_{G_s}^H V_s$, where $\mathscr{S}$ is a complete set of representatives of the set $H\backslash G/K$ of all H-K double cosets in G, $G_s = H \cap sKs^{-1}$, and (σ_s, V_s) is defined by setting $V_s = V$ and $\sigma_s(x) = \sigma(s^{-1}xs)$ for all $s \in \mathscr{S}$ and $x \in G_s$.

Now, Theorem 5.2 may be seen as a concrete realization of Mackey's Lemma. Indeed, by Lemma 5.1, we have the decomposition $G = CB \equiv C1_G B$ and $C \cap B = Z$ (i.e., in our previous notation: $H = C$, $K = B$, $\mathscr{S} = \{1_G\}$, and $G_{1_G} = H \cap K = Z$, so that $(V_{1_G}, \sigma_{1_G}) = (V, \mathrm{Res}_Z^B \sigma)$ for all $\sigma \in \widehat{B}$). Therefore we have

$$\begin{aligned}
\mathrm{Res}_C^G \widehat{\chi}_{\psi_1,\psi_2} &= \mathrm{Res}_C^G \mathrm{Ind}_B^G \chi_{\psi_1,\psi_2} \\
\text{(by Mackey's lemma)} &= \mathrm{Ind}_Z^C \mathrm{Res}_Z^B \chi_{\psi_1,\psi_2} \\
&\sim \mathrm{Ind}_{\mathbb{F}_q^*}^{\mathbb{F}_{q^2}^*} \psi_1\psi_2 \qquad (5.21) \\
&= \bigoplus_{\substack{\nu\in\widehat{\mathbb{F}_{q^2}^*}:\\ \nu^\sharp=\psi_1\psi_2}} \nu,
\end{aligned}$$

compare with (5.19). Note also that (5.21) or, equivalently, its concrete realization (5.18) in Theorem 5.2, yields another proof of (5.14) (similarly, for (5.13)). This may be also interpreted as the first part of an alternative proof that C is a multiplicity-free subgroup of G; for the remaining part, involving cuspidal representations, see Theorem 5.4.

We now compute the parabolic spherical functions.

Theorem 5.3

(1) *The spherical function associated with the parabolic representation* $\widehat{\chi}_{\psi_1,\psi_2}$ *(where* $\psi_1, \psi_2 \in \widehat{\mathbb{F}_q^*}$ *satisfy that* $\nu_0^\sharp = \psi_1\psi_2$*) is given by (cf. Lemma 5.1.(1)):*

$$\phi^{\psi_1,\psi_2}\left[\begin{pmatrix} x & y \\ 0 & 1 \end{pmatrix}\begin{pmatrix} a & \eta b \\ b & a \end{pmatrix}\right] = \frac{\overline{\nu_0(a+ib)}}{q+1}\left[\sum_{\gamma\in\mathbb{F}_q} \nu_0(\gamma+i)\overline{\nu_0(x\gamma+y+i)} \cdot\psi_2\left(\frac{(x\gamma+y)^2-\eta}{x(\gamma^2-\eta)}\right) + \overline{\psi_1(x)}\right].$$

(2) *The spherical function associated with* $\widehat{\chi}^1_{\psi}$ *(where* $\psi \in \widehat{\mathbb{F}_q^*}$ *satisfies that* $\nu_0^\sharp = \psi^2$*) is given by (cf. Lemma 5.1(1)):*

$$\phi^{\psi}\left[\begin{pmatrix} x & y \\ 0 & 1 \end{pmatrix}\begin{pmatrix} a & \eta b \\ b & a \end{pmatrix}\right] = \frac{\overline{\nu_0(a+ib)}}{q+1}\left[\sum_{\gamma\in\mathbb{F}_q} \nu_0(\gamma+i)\overline{\nu_0(x\gamma+y+i)} \cdot\psi\left(\frac{(x\gamma+y)^2-\eta}{x(\gamma^2-\eta)}\right) + \overline{\psi(x)}\right].$$

Proof Again, we just prove (1). By virtue of Theorem 5.2 and (3.24) the spherical function ϕ^{ψ_1,ψ_2} is given by

$$\phi^{\psi_1,\psi_2}(g) = \frac{1}{q+1}\langle F_{\nu_0}, \lambda(g)F_{\nu_0}\rangle_{V_{\widehat{\chi}_{\psi_1,\psi_2}}}$$

for all $g \in G$. Thus, taking $g = \begin{pmatrix} x & y \\ 0 & 1 \end{pmatrix}\begin{pmatrix} a & \eta b \\ b & a \end{pmatrix}$, by (5.16) we have:

$$\begin{aligned}\phi^{\psi_1,\psi_2}\left[\begin{pmatrix} x & y \\ 0 & 1 \end{pmatrix}\begin{pmatrix} a & \eta b \\ b & a \end{pmatrix}\right] &= \frac{\overline{\nu_0(a+ib)}}{q+1}\left\langle \lambda\begin{pmatrix} x & y \\ 0 & 1 \end{pmatrix}^{-1} F_{\nu_0}, F_{\nu_0}\right\rangle \\ \text{(by (1.24) and (5.20))} &= \frac{\overline{\nu_0(a+ib)}}{q^2-1}\sum_{\alpha+i\beta\in\mathbb{F}_{q^2}^*} F_{\nu_0}\left[\begin{pmatrix} x & y \\ 0 & 1 \end{pmatrix}\begin{pmatrix} \alpha & \eta\beta \\ \beta & \alpha \end{pmatrix}\right]\overline{F_{\nu_0}\begin{pmatrix} \alpha & \eta\beta \\ \beta & \alpha \end{pmatrix}}\end{aligned}$$

$$
\begin{aligned}
(\text{by } (5.1)) &= \frac{\overline{\nu_0(a+ib)}}{q^2-1}\left(\sum_{\substack{\alpha+i\beta\in\mathbb{F}_{q^2}^*:\\ \beta\neq 0}} F_{\nu_0}\left[\begin{pmatrix} u & \eta v\\ v & u\end{pmatrix}\begin{pmatrix} a & b\\ 0 & 1\end{pmatrix}\right]\nu_0(\alpha+i\beta)\right.\\
&\quad\left.+\sum_{\alpha\in\mathbb{F}_q^*} F_{\nu_0}\left[\begin{pmatrix} x & y\\ 0 & 1\end{pmatrix}\begin{pmatrix} \alpha & 0\\ 0 & \alpha\end{pmatrix}\right]\nu_0(\alpha)\right)\\
(\text{by } (5.16)) &= \frac{\overline{\nu_0(a+ib)}}{q^2-1}\left(\sum_{\substack{\alpha+i\beta\in\mathbb{F}_{q^2}^*:\\ \beta\neq 0}} \overline{\nu_0(u+iv)\psi_1(a)}\nu_0(\alpha+i\beta)\right.\\
&\quad\left.+\sum_{\alpha\in\mathbb{F}_q^*} \overline{\nu_0(\alpha)\psi_1(x)}\nu_0(\alpha)\right)\\
(\text{by } (5.2) \text{ with } \gamma=\alpha\beta^{-1}) &= \frac{\overline{\nu_0(a+ib)}}{q+1}\left(\sum_{\gamma\in\mathbb{F}_q}\nu_0(\gamma+i)\overline{\nu_0(x\gamma+y+i)}\right.\\
&\quad\left.\cdot\overline{\nu_0\left(\frac{x(\gamma^2-\eta)}{(x\gamma+y)^2-\eta}\right)}\,\overline{\psi_1\left(\frac{(x\gamma+y)^2-\eta}{x(\gamma^2-\eta)}\right)}+\overline{\psi_1(x)}\right)\\
(\nu^\sharp=\psi_1\psi_2) &= \frac{\overline{\nu_0(a+ib)}}{q+1}\left(\sum_{\gamma\in\mathbb{F}_q}\nu_0(\gamma+i)\overline{\nu_0(x\gamma+y+i)}\cdot\right.\\
&\quad\left.\cdot\psi_2\left(\frac{(x\gamma+y)^2-\eta}{x(\gamma^2-\eta)}\right)+\overline{\psi_1(x)}\right).
\end{aligned}
$$

□

5.6 Spherical Functions for $(\mathrm{GL}(2, \mathbb{F}_q), C, \nu_0)$: the Cuspidal Case

We now examine, more closely, the restriction from $G = \mathrm{GL}(2, \mathbb{F}_q)$ to C of a cuspidal representation. In comparison with the parabolic case examined above, we follow a slightly different approach, because cuspidal representations are quite intractable. We use the standard notation $(\chi, j, \dots)$ as in Sect. 5.3.

Lemma 5.2 *Let $\nu, \nu_0 \in \widehat{\mathbb{F}^*_{q^2}}$ and suppose that $\nu^\sharp = \nu_0^\sharp$. Then*

$$\sum_{\gamma \in \mathbb{F}_q} \nu(\gamma + i)\overline{\nu_0(\gamma + i)} = (q+1)\delta_{\nu,\nu_0} - 1.$$

Proof

$$\begin{aligned}
\sum_{\gamma \in \mathbb{F}_q} \nu(\gamma + i)\overline{\nu_0(\gamma + i)} &= \frac{1}{q-1} \sum_{\substack{\alpha \in \mathbb{F}_q \\ \beta \in \mathbb{F}_q^*}} \nu(\alpha\beta^{-1} + i)\overline{\nu_0(\alpha\beta^{-1} + i)} \\
(\nu^\sharp = \nu_0^\sharp) \quad &= \frac{1}{q-1} \sum_{\substack{a+i\beta \in \mathbb{F}^*_{q^2}: \\ \beta \neq 0}} \nu(\alpha + i\beta)\overline{\nu_0(\alpha + i\beta)} \\
&= \frac{1}{q-1} \sum_{w \in \mathbb{F}^*_{q^2}} \nu(w)\overline{\nu_0(w)} - \frac{1}{q-1} \sum_{\alpha \in \mathbb{F}_q^*} \nu(\alpha)\overline{\nu_0(\alpha)} \\
(\nu^\sharp = \nu_0^\sharp) \quad &= (q+1)\delta_{\nu,\nu_0} - 1
\end{aligned}$$

□

Theorem 5.4 *Let $\nu, \nu_0 \in \widehat{\mathbb{F}^*_{q^2}}$ and suppose that ν is indecomposable. Then the orthogonal projection E_{ν_0} onto the ν_0-isotypic component of $\mathrm{Res}^G_C \rho_\nu$ is given by:*

$$[E_{\nu_0} f](y) = \sum_{x \in \mathbb{F}_q^*} f(x) F_0(x, y)$$

for all $y \in \mathbb{F}_q^*$ and $f \in L(F_q^*)$, where

$$\begin{aligned}
F_0(x, y) = \frac{\delta_{\nu_0^\sharp, \nu^\sharp}}{q+1} \Bigg[&-\nu(-x) \sum_{\gamma \in \mathbb{F}_q} \overline{\nu_0(\gamma + i)}\chi(\gamma(x^{-1} + y^{-1}) \\
&\cdot j(x^{-1}y^{-1}(\gamma^2 - \eta)) + \delta_x(y) \Bigg].
\end{aligned} \tag{5.22}$$

Moreover:

- *if $\nu_0^\sharp \neq \nu^\sharp$ then $F_0 \equiv 0$ and $E_{\nu_0} \equiv 0$;*
- *if $\nu_0^\sharp = \nu^\sharp$, but $\nu_0 = \nu$ or $\nu_0 = \overline{\nu}$, then again $E_{\nu_0} = 0$;*
- *if $\nu_0^\sharp = \nu^\sharp$ and $\nu_0 \neq \nu, \overline{\nu}$ then E_{ν_0} is a one-dimensional projection; in particular, $tr(E_{\nu_0}) = 1$, that is, $\sum_{x \in \mathbb{F}_q^*} F_0(x, x) = 1$.*

Proof Note that all itemized statements may be deduced from (5.15) (clearly, not the formula for F_0), but we prefer to give another proof of these basic facts in order to check the validity of the formula for E_{ν_0} (which is quite cumbersone). From the projection formula (1.18) (with G therein now equal to C) and the explicit expressions of ρ_ν (cf. (5.9) and (5.10)), we deduce that, for $f \in L(\mathbb{F}_q^*)$ and $y \in \mathbb{F}_q^*$,

$$
\begin{aligned}
[E_{\nu_0} f](y) &= \frac{1}{|C|}\sum_{g\in C}\overline{\nu_0(g)}[\rho_\nu(g)f](y)\\
&= \frac{1}{q^2-1}\Bigg[-\sum_{\substack{\alpha+i\beta\in\mathbb{F}_{q^2}^*:\\ \beta\neq 0}}\overline{\nu_0(\alpha+i\beta)}\cdot\sum_{x\in\mathbb{F}_q^*}\nu(-\beta x)\chi(\alpha\beta^{-1}y^{-1}+\beta^{-1}\alpha x^{-1})\\
&\qquad\cdot j(\beta^{-2}y^{-1}x^{-1}(\alpha^2-\eta\beta^2))f(x)+\sum_{\alpha\in\mathbb{F}_q^*}\overline{\nu_0(\alpha)}\nu(\alpha)f(y)\Bigg]\\
(\gamma=\alpha\beta^{-1}) &= \frac{1}{q^2-1}\Bigg\{-\sum_{\gamma\in\mathbb{F}_q}\overline{\nu_0(\gamma+i)}\Bigg[\sum_{\beta\in\mathbb{F}_q^*}\overline{\nu_0(\beta)}\nu(\beta)\Bigg]\\
&\qquad\cdot\sum_{x\in\mathbb{F}_q^*}\nu(-x)\chi(\gamma(x^{-1}+y^{-1}))j(x^{-1}y^{-1}(\gamma^2-\eta))f(x)\\
&\qquad+\delta_{\nu_0^\sharp,\nu^\sharp}(q-1)f(y)\Bigg\}\\
&= \frac{\delta_{\nu_0^\sharp,\nu^\sharp}}{q+1}\sum_{x\in\mathbb{F}_q^*}\Bigg[-\nu(-x)\sum_{\gamma\in\mathbb{F}_q}\overline{\nu_0(\gamma+i)}\\
&\qquad\cdot\chi(\gamma(x^{-1}+y^{-1}))j(x^{-1}y^{-1}(\gamma^2-\eta))+\delta_x(y)\Bigg]f(x),
\end{aligned}
\tag{5.23}
$$

where in the last identity we have written $f(y)=\sum_{x\in\mathbb{F}_q^*}\delta_x(y)f(x)$. This gives the expression for E_{ν_0}. In particular, $E_{\nu_0}=0$ if $\nu_0^\sharp\neq\nu^\sharp$. We now compute the trace of E_{ν_0} (assuming $\nu_0^\sharp=\nu^\sharp$):

$$
\begin{aligned}
\mathrm{tr}(E_{\nu_0}) &= \sum_{x\in\mathbb{F}_q^*}F_0(x,x)\\
&= \frac{1}{q+1}\Bigg[\sum_{x\in\mathbb{F}_q^*}-\nu(-x)\sum_{\gamma\in\mathbb{F}_q}\overline{\nu_0(\gamma+i)}\chi(2\gamma x^{-1})\\
&\qquad\cdot j(x^{-2}(\gamma^2-\eta))+(q-1)\Bigg]
\end{aligned}
$$

$$= \frac{1}{q+1}\left[\sum_{\gamma\in\mathbb{F}_q} -\overline{\nu_0(\gamma+i)} \sum_{x\in\mathbb{F}_q^*} \nu(-x)\chi(2\gamma x^{-1}) \cdot j(x^{-2}(\gamma^2-\eta)) + (q-1)\right]$$

$$\text{(by (5.7))} \quad = \frac{1}{q+1}\left[\sum_{\gamma\in\mathbb{F}_q} -\overline{\nu_0(\gamma+i)}\,[\nu(\gamma+i)+\overline{\nu}(\gamma+i)] + (q-1)\right]$$

$$\text{(by Lemma 5.2)} \quad = \frac{1}{q+1}\left[-(q+1)\delta_{\nu,\nu_0} - (q+1)\delta_{\overline{\nu},\nu_0} + 2 + (q-1)\right]$$

$$= \begin{cases} 0 & \text{if } \nu_0 = \nu \text{ or } \nu_0 = \overline{\nu} \\ 1 & \text{if } \nu_0 \neq \nu, \overline{\nu}. \end{cases}$$

□

From now on, we assume $\nu_0^\sharp = \nu^\sharp$ and $\nu_0 \neq \nu, \overline{\nu}$.

Remark 5.5 Since E_{ν_0} is the projection onto a *one-dimensional* subspace, there exists $f_0 \in L(\mathbb{F}_q^*)$ satisfying $\|f_0\| = 1$ such that

$$E_{\nu_0} f = \langle f, f_0 \rangle f_0,$$

that is,

$$F_0(x, y) = \overline{f_0(x)}\, f_0(y) \tag{5.24}$$

for all $x, y \in \mathbb{F}_q^*$, and

$$\rho_\nu(g) f_0 = \nu_0(g) f_0 \tag{5.25}$$

for all $g \in C$.

Note that, by virtue of (5.24), we have $f_0(y) = (\overline{f_0(1)})^{-1} F_0(1, y)$, where

$$F_0(1, y) = \frac{1}{q+1}\left[-\nu(-1)\sum_{\gamma\in\mathbb{F}_q} \overline{\nu_0(\gamma+i)}\chi(\gamma(1+y^{-1}))j(y^{-1}(\gamma^2-\eta)) + \delta_1(y)\right],$$

and, moreover,

$$|f_0(1)|^2 \equiv F(1,1) = \frac{1}{q+1}\left[-\nu(-1)\sum_{\gamma\in\mathbb{F}_q} \overline{\nu_0(\gamma+i)}\chi(2\gamma)j(\gamma^2-\eta) + 1\right].$$

We were unable to find a simple expression for f_0 satisfying (5.24) (resp. for the norm of $F_0(1, y)$ and for the value $|f_0(1)|$). We leave it as an *open problem* to find such possible simple expressions. Fortunately, this does not constitute an obstruction for our subsequent computation of the spherical functions.

Finally, note that the computation $\sum_{x\in\mathbb{F}_q^*} F_0(x, x) = 1$ in Theorem 5.4 is equivalent to $\sum_{x\in\mathbb{F}_q^*} |f_0(x)|^2 = 1$.

A group theoretical proof of (5.25) is trivial: from the identity in the first line of (5.23) we get, for $g \in C$ and $f \in L(\mathbb{F}_q^*)$,

$$\begin{aligned} \rho_\nu(g)E_{\nu_0} f &= \frac{1}{|C|}\sum_{g'\in C} \overline{\nu_0(g')}\rho_\nu(gg')f \\ (\text{set } h = gg') \quad &= \frac{1}{|C|}\sum_{h\in C} \overline{\nu_0(g^{-1}h)}\rho_\nu(h)f \\ &= \nu_0(g)E_{\nu_0} f. \end{aligned} \tag{5.26}$$

Then (5.25) follows from (5.26) after observing that $f_0 = E_{\nu_0} f_0$. Note also that (5.26) relies on the identity $\rho_\nu(g)\rho_\nu(g') = \rho_\nu(gg')$, a quite nontrivial fact that has been proved analytically in [17, Section 14.6].

In the following, we give a direct analytic proof of (5.25), in order to also check the validity of our formulas (and in view of the open problem in Remark 5.5 and the intractability of the expression for cuspidal representations). We use the notation in (5.24) (but we always use $F_0(x, y)$ in the computation).

We need a preliminary, quite useful and powerful lemma, which is the core of our analytical computations. For all $x, y \in \mathbb{F}_q^*$ it is convenient to set

$$\widetilde{F_0}(x, y) = -\nu(-x)\sum_{\gamma\in\mathbb{F}_q} \overline{\nu_0(\gamma + i)}\chi(\gamma(x^{-1}+y^{-1}))j(x^{-1}y^{-1}(\gamma^2-\eta)) \tag{5.27}$$

so that

$$F_0(x, y) = \frac{1}{q+1}[\widetilde{F_0}(x, y) + \delta_x(y)]. \tag{5.28}$$

Lemma 5.3 *With the notation and the assumptions in Theorem 5.4, we have*

$$\begin{aligned} &\sum_{z\in\mathbb{F}_q^*} \widetilde{F_0}(x, z)\nu(-z)\chi(\delta(y^{-1} + z^{-1}))j(y^{-1}z^{-1}(\delta^2 - \eta)) \\ &\quad = -\nu_0(\delta + i)[\widetilde{F_0}(x, y) + \delta_x(y)] - \nu(-x)\chi(\delta(x^{-1} + y^{-1}))j(x^{-1}y^{-1}(\delta^2 - \eta)) \end{aligned} \tag{5.29}$$

and

$$\sum_{z\in\mathbb{F}_q^*} \widetilde{F}_0(x, z)\widetilde{F}_0(z, y) = (q-1)\widetilde{F}_0(x, y) + q\delta_x(y), \tag{5.30}$$

for all $\delta \in \mathbb{F}_q$ *and* $x, y \in \mathbb{F}_q^*$.

Proof The left hand side of (5.29) is equal to

$$\begin{aligned}
&\sum_{z\in\mathbb{F}_q^*}\left[-\nu(-x)\sum_{\gamma\in\mathbb{F}_q}\overline{\nu_0(\gamma+i)\chi(\gamma(x^{-1}+z^{-1}))j(x^{-1}z^{-1}(\gamma^2-\eta))}\right]\\
&\quad\cdot\left[\nu(-z)\chi(\delta(y^{-1}+z^{-1}))j(y^{-1}z^{-1}(\delta^2-\eta))\right]\\
&= -\nu(x)\nu_0(\delta+i)\sum_{\gamma\in\mathbb{F}_q}\overline{\nu_0((\gamma+i)(\delta+i))\chi(\gamma x^{-1}+\delta y^{-1})}\\
&\quad\cdot\sum_{z\in\mathbb{F}_q^*}\chi((\gamma+\delta)z^{-1})\nu(z)j(z^{-1}x^{-1}(\gamma^2-\eta))j(z^{-1}y^{-1}(\delta^2-\eta))\\
&=_* -\nu(x)\nu_0(\delta+i)\sum_{\substack{\gamma\in\mathbb{F}_q:\\ \gamma\neq-\delta}}\overline{\nu_0((\gamma\delta+\eta)+i(\gamma+\delta))\chi(\gamma x^{-1}+\delta y^{-1})}\cdot\nu(\gamma+\delta)\\
&\quad\cdot\sum_{t\in\mathbb{F}_q^*}\chi(t)\nu(t^{-1})j(t(\gamma+\delta)^{-1}x^{-1}(\gamma^2-\eta))j(t(\gamma+\delta)^{-1}y^{-1}(\delta^2-\eta))\\
&\quad-\nu(x)\nu_0(\delta+i)\overline{\nu_0(-\delta^2+\eta)}\chi(\delta y^{-1}-\delta x^{-1})\\
&\quad\cdot\sum_{z\in\mathbb{F}_q^*}\nu(z)j(z^{-1}x^{-1}(\delta^2-\eta))j(z^{-1}y^{-1}(\delta^2-\eta))\\
&=_{**} \nu(x)\nu_0(\delta+i)\sum_{\substack{\gamma\in\mathbb{F}_q:\\ \gamma\neq-\delta}}\overline{\nu_0((\gamma\delta+\eta)+i(\gamma+\delta))\cdot\nu(\gamma+\delta)\cdot\chi(\gamma x^{-1}+\delta y^{-1})}\\
&\quad\cdot\chi(-(\gamma+\delta)^{-1}[x^{-1}(\gamma^2-\eta)+y^{-1}(\delta^2-\eta)])\\
&\quad\cdot\nu(-1)j((\gamma+\delta)^{-2}x^{-1}y^{-1}(\delta^2-\eta)(\gamma^2-\eta))\\
&\quad-\nu(x)\nu_0(\delta+i)\overline{\nu_0(-\delta^2+\eta)}\chi(\delta y^{-1}-\delta x^{-1})\delta_x(y)\nu(-x^{-1}(\delta^2-\eta)),
\end{aligned}$$

where $=_*$ follows after taking $t = (\gamma+\delta)z^{-1}$ and $=_{**}$ follows from (5.6) and (5.5). To complete our calculations will use the following elementary algebraic identities. Let us first set, for $\gamma \neq -\delta$,

$$\epsilon = \frac{\gamma\delta+\eta}{\gamma+\delta}.$$

Then, we have:

- $\overline{\nu_0((\gamma\delta+\eta)+i(\gamma+\delta))}\cdot\nu(\gamma+\delta)=\overline{\nu_0(\epsilon+i)}$ (recall that $\nu^\sharp=\nu_0^\sharp$);
- $\gamma=\frac{\delta\epsilon-\eta}{\delta-\epsilon}$ so that $\sum_{\gamma\neq-\delta}$ is equivalent to $\sum_{\epsilon\neq\delta}$;
- $\gamma-(\gamma+\delta)^{-1}(\gamma^2-\eta)=\frac{\gamma\delta+\eta}{\gamma+\delta}=\epsilon=\delta-(\gamma+\delta)^{-1}(\delta^2-\eta)$ (this is the coefficient of both x^{-1} and y^{-1} in the grouped argument of χ);
- in the argument of j:

$$
\begin{aligned}
(\gamma+\delta)^{-2}(\gamma^2-\eta)(\delta^2-\eta) &= (\gamma+\delta)^{-2}(\gamma+i)(\delta+i)(\gamma-i)(\delta-i)\\
&= (\gamma+\delta)^{-2}[(\gamma\delta+\eta)+(\gamma+\delta)i]\\
&\quad\cdot[(\gamma\delta+\eta)-(\gamma+\delta)i)]\\
&= \left(\frac{\gamma\delta+\eta}{\gamma+\delta}\right)^2-\eta=\epsilon^2-\eta;
\end{aligned}
$$

- $\nu(x)\overline{\nu_0(-\delta^2+\eta)}\nu(-x^{-1}(\delta^2-\eta))=1$;
- $\chi(\delta y^{-1}-\delta x^{-1})\delta_x(y)=\delta_x(y)$.

Therefore, continuing the above calculations, the left hand side of (5.29) equals

$$
\begin{aligned}
&\nu(-x)\nu_0(\delta+i)\sum_{\substack{\epsilon\in\mathbb{F}_q:\\ \epsilon\neq\delta}}\overline{\nu_0(\epsilon+i)}\chi(\epsilon(x^{-1}+y^{-1}))\\
&\quad\cdot j(x^{-1}y^{-1}(\epsilon^2-\eta))-\nu_0(\delta+i)\delta_x(y)\\
&= -\nu_0(\delta+i)\Big[\widetilde{F}_0(x,y)+\delta_x(y)+\nu(-x)\overline{\nu_0(\delta+i)}\chi(\delta(x^{-1}+y^{-1}))\\
&\quad\cdot j(x^{-1}y^{-1}(\delta^2-\eta))\Big]\\
&= -\nu_0(\delta+i)\left[\widetilde{F}_0(x,y)+\delta_x(y)\right]-\nu(-x)\chi(\delta(x^{-1}+y^{-1}))j(x^{-1}y^{-1}(\delta^2-\eta)).
\end{aligned}
$$

This completes the proof of (5.29). Finally, (5.30) follows from

$$
\begin{aligned}
\sum_{z\in\mathbb{F}_q^*}\widetilde{F}_0(x,z)\widetilde{F}_0(z,y) &= -\sum_{\delta\in\mathbb{F}_q}\overline{\nu_0(\delta+i)}\sum_{z\in\mathbb{F}_q^*}\widetilde{F}_0(x,z)\nu(-z)\chi(\delta(y^{-1}+z^{-1}))\\
&\quad\cdot j(y^{-1}z^{-1}(\delta^2-\eta))\\
&= (q-1)\widetilde{F}_0(x,y)+q\delta_x(y),
\end{aligned}
$$

where the last equality follows from (5.29) and (5.27). □

***Proof (Analytic Proof of* (5.25))** Let $g = \begin{pmatrix} \alpha & \eta\beta \\ \beta & \alpha \end{pmatrix} \in C$ and $x, z \in \mathbb{F}_q^*$. First of all, note that if $\beta = 0$ there is nothing to prove: from (5.9) it follows that

$$\rho_\nu \begin{pmatrix} \alpha & 0 \\ 0 & \alpha \end{pmatrix} f = \nu(\alpha) f = \nu_0(\alpha) f$$

for all $f \in L(\mathbb{F}_q^*)$, where the last equality follows from our assumption $\nu_0^\sharp = \nu^\sharp$.

Suppose now that $\beta \neq 0$. Since $f_0 \not\equiv 0$, we can find $x \in \mathbb{F}_q^*$ such that $f_0(x) \neq 0$. Taking into account (5.10), we then have

$$\begin{aligned}
\overline{f_0(x)}[\rho_\nu(g) f_0](y) &= -\overline{f_0(x)} \sum_{z \in \mathbb{F}_q^*} \nu(-\beta z)\chi(\alpha\beta^{-1}y^{-1} + \alpha\beta^{-1}z^{-1}) \\
&\quad \cdot j(\beta^{-2}y^{-1}z^{-1}(\alpha^2 - \eta\beta^2)) f_0(z) \\
\text{(by (5.24))} &= -\sum_{z \in \mathbb{F}_q^*} \nu(-\beta z)\chi(\alpha\beta^{-1}y^{-1} + \alpha\beta^{-1}z^{-1}) \\
&\quad \cdot j(\beta^{-2}y^{-1}z^{-1}(\alpha^2 - \eta\beta^2)) F_0(x, z) \\
\text{(by (5.28) and } \delta = \alpha\beta^{-1}) &= -\frac{\nu(\beta)}{q+1} \sum_{z \in \mathbb{F}_q^*} \nu(-z)\chi(\delta(y^{-1} + z^{-1})) \\
&\quad \cdot j(y^{-1}z^{-1}(\delta^2 - \eta))\widetilde{F}_0(x, z) \\
&\quad - \frac{\nu(\beta)}{q+1} \sum_{z \in \mathbb{F}_q^*} \nu(-z)\chi(\delta(y^{-1} + z^{-1})) \\
&\quad \cdot j(y^{-1}z^{-1}(\delta^2 - \eta))\delta_x(z) \\
\text{(by (5.29))} &= \frac{\nu(\beta)}{q+1} \nu_0(\alpha\beta^{-1} + i)[\widetilde{F}_0(x, y) + \delta_x(y)] \\
&\quad + \frac{\nu(\beta)}{q+1} \nu(-x)\chi(\delta(x^{-1} + y^{-1})) j(x^{-1}y^{-1}(\delta^2 - \eta)) \\
&\quad - \frac{\nu(\beta)}{q+1} \nu(-x)\chi(\delta(y^{-1} + x^{-1})) j(y^{-1}x^{-1}(\delta^2 - \eta)) \\
\text{(by (5.28) and } \nu_0^\sharp = \nu^\sharp) &= \nu_0(\alpha + i\beta) F_0(x, y) \\
\text{(by (5.24))} &= \overline{f_0(x)} \nu_0(\alpha + i\beta) f_0(y).
\end{aligned}$$

After simplifying (recall that $f_0(x) \neq 0$), one immediately deduces (5.25). □

We now want to show how Lemma 5.3 can be used to derive by means of purely *analytical* methods the other basic properties of the matrix $F_0(x, y)$ (recall that we have already proved, analytically, that its trace is equal to 1 (cf. in Theorem 5.4)).

Theorem 5.5 *The function $F_0(x, y) = \overline{f_0(x)} f_0(y)$, $x, y \in \mathbb{F}_q^*$ (cf. (5.24)), satisfies the following identities:*

$$\sum_{z \in \mathbb{F}_q^*} F_0(x, z) F_0(z, y) = F_0(x, y), \quad \textit{(idempotence)} \tag{5.31}$$

$$\overline{F_0(x, y)} = F_0(y, x), \quad \textit{(self-adjointness)} \tag{5.32}$$

for all $x, y, z \in \mathbb{F}_q^*$.

Proof By (5.28), we have

$$\begin{aligned}
\sum_{z \in \mathbb{F}_q^*} F_0(x, z) F_0(z, y) &= \frac{1}{(q+1)^2} \sum_{z \in \mathbb{F}_q^*} [\widetilde{F}_0(x, z) + \delta_x(z)] \cdot [\widetilde{F}_0(z, y) + \delta_z(y)] \\
&= \frac{1}{(q+1)^2} \left[\sum_{z \in \mathbb{F}_q^*} \widetilde{F}_0(x, z) \widetilde{F}_0(z, y) + 2\widetilde{F}_0(x, y) + \delta_x(y) \right] \\
(\text{by (5.30)}) &= \frac{1}{(q+1)^2} \left[(q+1)\widetilde{F}_0(x, y) + (q+1)\delta_x(y) \right] \\
&= F_0(x, y)
\end{aligned}$$

proving (5.31).

From $\overline{\nu(w)} = \nu(w^{-1})$, $\overline{\chi(z)} = \chi(-z)$, (5.8), and (5.22) we deduce

$$\begin{aligned}
\overline{F_0(x, y)} &= \frac{1}{q+1} \left[-\nu(-x^{-1}) \sum_{\gamma \in \mathbb{F}_q} \nu_0(\gamma + i) \chi(-\gamma(x^{-1} + y^{-1})) \right. \\
&\qquad \left. \cdot j(x^{-1} y^{-1}(\gamma^2 - \eta)) \overline{\nu(-x^{-1} y^{-1}(\gamma^2 - \eta))} + \delta_x(y) \right] \\
(\nu_0^\sharp = \nu^\sharp) &= \frac{1}{q+1} \left[-\nu(-y) \sum_{\gamma \in \mathbb{F}_q} \nu_0(\gamma + i) \overline{\nu_0(\gamma + i) \nu_0(\gamma - i)} \nu_0(-1) \right. \\
&\qquad \left. \cdot \chi(-\gamma(x^{-1} + y^{-1})) j(x^{-1} y^{-1}(\gamma^2 - \eta)) + \delta_x(y) \right]
\end{aligned}$$

$$(\gamma \mapsto -\gamma) = \frac{1}{q+1}\left[-\nu(-y)\sum_{\gamma\in\mathbb{F}_q}\overline{\nu_0(\gamma+i)}\chi(\gamma(x^{-1}+y^{-1}))\right.$$
$$\left.\cdot j(x^{-1}y^{-1}(\gamma^2-\eta))+\delta_x(y)\right]$$
$$(\text{by (5.22)}) = F_0(y, x).$$

□

We end our analytic verifications by proving the orthogonality relations for different projections (corresponding to different indecomposable characters of $\mathbb{F}^*_{q^2}$). Thus let $\mu_0 \in \widehat{\mathbb{F}^*_{q^2}}$ and suppose that $\mu_0^\sharp = \nu^\sharp$, but $\mu_0 \neq \nu_0, \nu, \overline{\nu}$. Define $G_0(x, y)$ and $\widetilde{G_0}(x, y)$ as $F_0(x, y)$ and $\widetilde{F_0}(x, y)$ in Theorem 5.4 and (5.27), respectively, but with μ_0 in place of ν_0, so that, as in (5.28),

$$G_0(x, y) = \frac{1}{q+1}\left[\widetilde{G_0}(x, y) + \delta_x(y)\right].$$

Proposition 5.2 *With the above notation, we have*

$$\sum_{z\in\mathbb{F}_q^*} F_0(x, z)G_0(z, y) = 0 \tag{5.33}$$

for all $x, y \in \mathbb{F}_q^*$.

Proof Arguing as in the proof of the second identity in Lemma 5.3 we have

$$\sum_{z\in\mathbb{F}_q^*}\widetilde{F_0}(x, z)\widetilde{G_0}(z, y) = -\sum_{\delta\in\mathbb{F}_q}\overline{\mu_0(\delta+i)}\sum_{z\in\mathbb{F}_q^*}\widetilde{F_0}(x, z)\nu(-z)$$
$$\cdot\chi(\delta(y^{-1}+z^{-1}))j(y^{-1}z^{-1}(\delta^2-\eta))$$
$$(\text{by (5.29) and (5.27) for } \widetilde{G_0}) = \sum_{\delta\in\mathbb{F}_q}\nu_0(\delta+i)\overline{\mu_0(\delta+i)}[\widetilde{F_0}(x, y)+\delta_x(y)] - \widetilde{G_0}(x, y)$$
$$(\text{by Lemma 5.2}) = -\widetilde{F_0}(x, y) - \widetilde{G_0}(x, y) - \delta_x(y).$$

Therefore,

$$\sum_{z\in\mathbb{F}_q^*}F_0(x, z)G_0(z, y) = \frac{1}{(q+1)^2}\sum_{z\in\mathbb{F}_q^*}\left([\widetilde{F_0}(x, z)+\delta_x(z)]\cdot[\widetilde{G_0}(z, y)+\delta_y(z)]\right)$$
$$= \frac{1}{(q+1)^2}\left(\sum_{z\in\mathbb{F}_q^*}\widetilde{F_0}(x, z)\widetilde{G_0}(z, y)+\widetilde{F_0}(x, y)+\widetilde{G_0}(x, y)+\delta_x(y)\right)$$

$$= \frac{-\widetilde{F_0}(x,y) - \widetilde{G_0}(x,y) - \delta_x(y) + \widetilde{F_0}(x,y) + \widetilde{G_0}(x,y) + \delta_x(y)}{(q+1)^2}$$
$$= 0,$$

where the last but one equality follows from the previous computations. □

We can now state and prove the analogue of Theorem 5.3 for cuspidal representations, completing the computation of the corresponding spherical functions.

Theorem 5.6 *Let $\nu_0, \nu \in \widehat{\mathbb{F}^*_{q^2}}$ indecomposable and suppose that $\nu^\sharp = \nu_0^\sharp$, but $\nu, \overline{\nu} \neq \nu_0$. Then the spherical function of the multiplicity-free triple (G, C, ν_0) associated with the cuspidal representation ρ_ν is given by (cf. Lemma 5.1(1)):*

$$\phi^\nu\left[\begin{pmatrix} x & y \\ 0 & 1\end{pmatrix}\begin{pmatrix} a & \eta b \\ b & a\end{pmatrix}\right] = -\frac{\overline{\nu_0(a+ib)}}{q+1}\sum_{z\in\mathbb{F}_q^*}\nu(-x^{-1}z)\chi(-yz^{-1})$$
$$\cdot \sum_{\gamma\in\mathbb{F}_q}\overline{\nu_0(\gamma+i)}\chi(\gamma z^{-1}(x+1))j(xz^{-2}(\gamma^2-\eta))$$
$$+\frac{\overline{\nu_0(a+ib)}}{q+1}\delta_{x,1}(q\delta_{y,0}-1).$$

Proof For all $g \in G$, taking into account (3.24), we have

$$\phi^\nu(g) = \langle f_0, \rho_\nu(g) f_0\rangle_{L(\mathbb{F}_q^*)} = \sum_{z\in\mathbb{F}_q^*} f_0(z)\overline{[\rho_\nu(g) f_0](z)},$$

so that, writing $g = \begin{pmatrix} x & y \\ 0 & 1\end{pmatrix}\begin{pmatrix} a & \eta b \\ b & a\end{pmatrix}$, we have

$$\phi^\nu(g) = \sum_{z\in\mathbb{F}_q^*} f_0(z)\overline{\left[\rho_\nu\begin{pmatrix} x & y \\ 0 & 1\end{pmatrix}\rho_\nu\begin{pmatrix} a & \eta b \\ b & a\end{pmatrix} f_0\right](z)}$$
$$\text{(by (5.25))} \quad = \overline{\nu_0(a+ib)}\sum_{z\in\mathbb{F}_q^*} f_0(z)\overline{\left[\rho_\nu\begin{pmatrix} x & y \\ 0 & 1\end{pmatrix} f_0\right](z)}$$
$$\text{(by (5.9))} = \overline{\nu_0(a+ib)}\sum_{z\in\mathbb{F}_q^*}\chi(-yz^{-1}) f_0(z)\overline{f_0(x^{-1}z)}$$

$$\text{(by (5.22) and (5.24))} = \overline{\nu_0(a+ib)} \sum_{z\in\mathbb{F}_q^*} \chi(-yz^{-1}) \frac{1}{q+1} \Big[-\nu(-x^{-1}z) \cdot \sum_{\gamma\in\mathbb{F}_q} \overline{\nu_0(\gamma+i)} \chi(\gamma z^{-1}(1+x)) \cdot j(xz^{-2}(\gamma^2-\eta)) + \delta_{x^{-1}z}(z) \Big].$$

Then we end the proof just by noticing that

$$\sum_{z\in\mathbb{F}_q^*} \chi(-yz^{-1})\delta_{x^{-1}z}(z) = \delta_{x,1} \sum_{z\in\mathbb{F}_q^*} \chi(-yz^{-1}) = \delta_{x,1}(q\delta_{y,0}-1). \qquad \square$$

Chapter 6
Harmonic Analysis of the Multiplicity-Free Triple $(\mathrm{GL}(2, \mathbb{F}_{q^2}), \mathrm{GL}(2, \mathbb{F}_q), \rho_\nu)$

In this chapter we study an example of a multiplicity-free triple where the representation that we induce has dimension greater than one.

Let $q = p^h$ with p an odd prime and $h \geq 1$. Set

$$G_1 = \mathrm{GL}(2, \mathbb{F}_q) \quad \text{and} \quad G_2 = \mathrm{GL}(2, \mathbb{F}_{q^2}).$$

Moreover (cf. Sect. 3.5), we denote by B_j (resp. U_j, resp. C_j) the Borel (resp. the unipotent, resp. the Cartan) subgroup of G_j, for $j = 1, 2$. Throughout this chapter, with the notation as in Sect. 5.3, we let $\nu \in \widehat{\mathbb{F}^*_{q^2}}$ be a fixed indecomposable character. We assume that $\nu^\sharp = \mathrm{Res}^{\mathbb{F}^*_{q^2}}_{\mathbb{F}^*_q} \nu$ is not a square: this slightly simplifies the decomposition into irreducibles. Finally, ρ_ν denotes the cuspidal representation of G_1 associated with ν.

Proposition 6.1 *(G_2, G_1, ρ_ν) is a multiplicity-free triple and*

$$\mathrm{Ind}^{G_2}_{G_1} \rho_\nu \sim \left(\bigoplus \widehat{\chi}_{\xi_1, \xi_2} \right) \oplus \left(\bigoplus \rho_\mu \right) \tag{6.1}$$

where

- *the first sum runs over all unordered pairs of distinct characters $\xi_1, \xi_2 \in \widehat{\mathbb{F}^*_{q^2}}$ such that $\mathrm{Res}^{\mathbb{F}^*_{q^2}}_{\mathbb{F}^*_q} \xi_1\xi_2 = \mathrm{Res}^{\mathbb{F}^*_{q^2}}_{\mathbb{F}^*_q} \nu$ but $\overline{\xi_1}\xi_2 \neq \nu, \overline{\nu}$;*
- *the second sum runs over all characters $\mu \in \mathbb{F}^*_{q^4}$ indecomposable over $\mathbb{F}^*_{q^2}$ such that $\mathrm{Res}^{\mathbb{F}^*_{q^4}}_{\mathbb{F}^*_q} \mu = \mathrm{Res}^{\mathbb{F}^*_{q^2}}_{\mathbb{F}^*_q} \nu$.*

Proof This is just an easy exercise. For instance, if χ^{ρ_ν} and χ^{ρ_μ} are the characters of ρ_ν and ρ_μ (with $\mu \in \widehat{\mathbb{F}^*_{q^4}}$ a generic indecomposable character), respectively, then by means of the character table of G_1 (cf. [17, Table 14.2]), the character table of the

T. Ceccherini-Silberstein et al., *Gelfand Triples and Their Hecke Algebras*, Lecture Notes in Mathematics 2267, https://doi.org/10.1007/978-3-030-51607-9_6

restrictions from G_2 to G_1 (cf. [17, Table 14.3]), and the table of conjugacy classes of G_1 ([17, Table 14.1]), we find:

$$\begin{aligned}
\frac{1}{|G_1|}\langle \chi^{\rho_\nu}, \mathrm{Res}^{G_2}_{G_1}\chi^{\rho_\mu}\rangle_{L(G_1)} &= \frac{(q^2-1)(q-1)}{|G_1|}\sum_{x\in\mathbb{F}_q^*}\nu(x)\overline{\mu(x)} \\
&\quad + \frac{q^2-1}{|G_1|}\sum_{x\in\mathbb{F}_q^*}\nu(x)\overline{\mu(x)} \\
&= \frac{(q^2-1)q}{|G_1|}\sum_{x\in\mathbb{F}_q^*}\nu(x)\overline{\mu(x)} \\
&= \begin{cases} 1 & \text{if } \mathrm{Res}^{\mathbb{F}_{q^4}^*}_{\mathbb{F}_q^*}\mu = \mathrm{Res}^{\mathbb{F}_{q^2}^*}_{\mathbb{F}_q^*}\nu \\ 0 & \text{otherwise,} \end{cases}
\end{aligned}$$

where the last equality follows from the orthogonality relations of characters. This yields the multiplicity of ρ_ν in $\mathrm{Res}^{G_2}_{G_1}\rho_\mu$ (by [17, Formula (10.17)]) and, in turn, the multiplicity of ρ_μ in $\mathrm{Ind}^{G_2}_{G_1}\rho_\nu$, by Frobenius reciprocity See [17, Section 14.10] for more computations of this kind. □

The result in the above proposition will complemented in Sect. A.2 where we shall study the induction of the trivial and the parabolic representations.

6.1 Spherical Functions for $(\mathrm{GL}(2, \mathbb{F}_{q^2}), \mathrm{GL}(2, \mathbb{F}_q), \rho_\nu)$: the Parabolic Case

We now compute the spherical functions associated with the parabolic representations in (6.1). In order to apply Mackey's lemma (cf. Remark 5.4), we need a preliminary result. Recall the definition of $i \in \mathbb{F}_{q^2}$ (cf. (3.31)).

Lemma 6.1 *Let* $W = \begin{pmatrix} i & 1 \\ 1 & 0 \end{pmatrix}$. *Then*

$$G_2 = G_1 B_2 \bigsqcup G_1 W B_2 \tag{6.2}$$

is the decomposition of G_2 *into* $G_1 - B_2$ *double cosets. Moreover,*

$$G_1 W B_2 = \left\{ \begin{pmatrix} a & b \\ c & d \end{pmatrix} \in G_2 : c \neq 0, ac^{-1} \notin \mathbb{F}_q \right\}. \tag{6.3}$$

Proof We prove the following facts for $\begin{pmatrix} a & b \\ c & d \end{pmatrix} \in G_2$:

(i) $\begin{pmatrix} a & b \\ c & d \end{pmatrix} \in G_1B_2 \Leftrightarrow c = 0$ or ($c \neq 0$ and $ac^{-1} \in \mathbb{F}_q$);

(ii) $\begin{pmatrix} a & b \\ c & d \end{pmatrix} \in G_1WB_2 \Leftrightarrow c \neq 0$ and $ac^{-1} \notin \mathbb{F}_q$.

Let $\begin{pmatrix} \alpha & \beta \\ \gamma & \delta \end{pmatrix} \in G_1$ and $\begin{pmatrix} x & y \\ 0 & z \end{pmatrix} \in B_2$. Then $\begin{pmatrix} \alpha & \beta \\ \gamma & \delta \end{pmatrix}\begin{pmatrix} x & y \\ 0 & z \end{pmatrix} = \begin{pmatrix} \alpha x & \alpha y + \beta z \\ \gamma x & \gamma y + \delta z \end{pmatrix}$ with either $\gamma x = 0$ or ($\gamma x \neq 0$ and $\alpha x(\gamma x)^{-1} = \alpha\gamma^{-1} \in \mathbb{F}_q$). Conversely, if $c \neq 0$ and $ac^{-1} \in \mathbb{F}_q$, then $\begin{pmatrix} a & b \\ c & d \end{pmatrix} = \begin{pmatrix} ac^{-1} & 1 \\ 1 & 0 \end{pmatrix}\begin{pmatrix} c & d \\ 0 & b - ad/c \end{pmatrix} \in G_1B_2$. The case $c = 0$ is trivial: indeed $B_2 \subseteq G_1B_2$. This shows (i).

We now consider

$$\begin{pmatrix} \alpha & \beta \\ \gamma & \delta \end{pmatrix}\begin{pmatrix} i & 1 \\ 1 & 0 \end{pmatrix}\begin{pmatrix} x & y \\ 0 & z \end{pmatrix} = \begin{pmatrix} (\alpha i + \beta)x & (\alpha i + \beta)y + \alpha z \\ (\gamma i + \delta)x & (\gamma i + \delta)y + \gamma z \end{pmatrix} \in G_1WB_2,$$

and $(\gamma, \delta) \neq (0, 0)$, $x \neq 0$ imply $(\gamma i + \delta)x \neq 0$, while $\det\begin{pmatrix} \alpha & \beta \\ \gamma & \delta \end{pmatrix} \neq 0$ implies $(\alpha i + \beta)(\gamma i + \delta)^{-1} \notin \mathbb{F}_q$. Conversely, if $c \neq 0$ and $ac^{-1} \notin \mathbb{F}_q$, then setting $ac^{-1} = \alpha i + \beta$, with $\alpha, \beta \in \mathbb{F}_q$, $\alpha \neq 0$, we have

$$\begin{pmatrix} a & b \\ c & d \end{pmatrix} = \begin{pmatrix} ac^{-1} & \alpha \\ 1 & 0 \end{pmatrix}\begin{pmatrix} c & d \\ 0 & (bc - ad)/(\alpha c) \end{pmatrix} = \begin{pmatrix} \alpha & \beta \\ 0 & 1 \end{pmatrix}\begin{pmatrix} i & 1 \\ 1 & 0 \end{pmatrix}\begin{pmatrix} c & d \\ 0 & (bc - ad)/(\alpha c) \end{pmatrix}. \tag{6.4}$$

Thus (ii) follows as well. □

Remark 6.1 Actually, we have a stronger result, which will be useful in the sequel, namely: any $\begin{pmatrix} a & b \\ c & d \end{pmatrix} \in G_2$ with $c \neq 0$ and $ac^{-1} \notin \mathbb{F}_q$ may be uniquely expressed as in (6.4).

Lemma 6.2 *In the decomposition* (6.2) *we have* $G_1 \cap WB_2W^{-1} = C_1$. *Moreover, for* $x_1, x_2 \in \mathbb{F}_q$ *we have:*

$$W^{-1}\begin{pmatrix} x_1 & \eta x_2 \\ x_2 & x_1 \end{pmatrix}W = \begin{pmatrix} x_1 + ix_2 & x_2 \\ 0 & x_1 - ix_2 \end{pmatrix}. \tag{6.5}$$

Proof Taking $x = x_1 + ix_2$, $z = z_1 + iz_2$ in $\mathbb{F}^*_{q^2}$ and $y = y_1 + iy_2 \in \mathbb{F}_{q^2}$, then

$$W\begin{pmatrix} x & y \\ 0 & z \end{pmatrix} W^{-1} = \begin{pmatrix} i & 1 \\ 1 & 0 \end{pmatrix}\begin{pmatrix} x_1+ix_2 & y_1+iy_2 \\ 0 & z_1+iz_2 \end{pmatrix}\begin{pmatrix} 0 & 1 \\ 1 & -i \end{pmatrix}$$
$$= \begin{pmatrix} \eta y_2 + z_1 + i(y_1+z_2) & \eta x_2 - \eta y_1 - \eta z_2 + i(x_1 - \eta y_2 - z_1) \\ y_1 + iy_2 & x_1 - \eta y_2 + i(x_2 - y_1) \end{pmatrix}$$

belongs to G_1 if and only if

$$\begin{array}{ll} y_1 + z_2 = 0 & x_1 - \eta y_2 - z_1 = 0 \\ y_2 = 0 & x_2 - y_1 = 0 \end{array}$$

equivalently,

$$y_2 = 0,\ z_2 = -y_1,\ y_1 = x_2, \text{ and } z_1 = x_1.$$

This is clearly the same as

$$W\begin{pmatrix} x & y \\ 0 & z \end{pmatrix} W^{-1} = \begin{pmatrix} x_1 & \eta x_2 \\ x_2 & x_1 \end{pmatrix} \in C_1.$$

The proof of (6.5) follows immediately from the above computations. □

In the following, as in Sect. 5.3, for $\xi \in \widehat{\mathbb{F}^*_{q^2}}$ we set $\xi^\sharp = \mathrm{Res}^{G_2}_{G_1}\xi$.

Theorem 6.1 *Let $\xi_1, \xi_2 \in \widehat{\mathbb{F}^*_{q^2}}$ with $\xi_1 \neq \xi_2, \overline{\xi_2}$. Then*

$$\mathrm{Res}^{G_2}_{G_1}\widehat{\chi}_{\xi_1,\xi_2} \sim \widehat{\chi}_{\xi_1^\sharp,\xi_2^\sharp} \oplus \mathrm{Ind}^{G_1}_{C_1}\xi_1\overline{\xi_2}. \tag{6.6}$$

Proof First of all, note that from (6.5) we have

$$\chi_{\xi_1,\xi_2}\left(W^{-1}\begin{pmatrix} x_1 & \eta x_2 \\ x_2 & x_1 \end{pmatrix} W\right) = \chi_{\xi_1,\xi_2}\begin{pmatrix} x_1+ix_2 & x_2 \\ 0 & x_1 - ix_2 \end{pmatrix}$$
$$(\text{by } (5.3)) = \xi_1(x_1+ix_2)\xi_2(x_1-ix_2)$$
$$= (\xi_1\overline{\xi_2})(x_1+ix_2).$$

In other words, $\mathrm{Res}^{B_2}_{W^{-1}C_1W}\chi_{\xi_1,\xi_2} \sim \xi_1\overline{\xi_2}$.

By Mackey's lemma and Lemma 6.2 (and $G_1 \cap B_2 = B_1$), we then have

$$\begin{aligned} \mathrm{Res}^{G_2}_{G_1}\widehat{\chi}_{\xi_1,\xi_2} &\sim \mathrm{Res}^{G_2}_{G_1}\mathrm{Ind}^{G_2}_{B_2}\chi_{\xi_1,\xi_2} \\ &\sim \mathrm{Ind}^{G_1}_{B_1}\chi_{\xi_1^\sharp,\xi_2^\sharp} \oplus \mathrm{Ind}^{G_1}_{C_1}\xi_1\overline{\xi_2} \\ &\sim \widehat{\chi}_{\xi_1^\sharp,\xi_2^\sharp} \oplus \mathrm{Ind}^{G_1}_{C_1}\xi_1\overline{\xi_2}. \end{aligned}$$

□

Remark 6.2 The above is a surprisingly and unexpected result: in the study of the parabolic representation involved in the triple (G_2, G_1, ρ_ν) we must use $\mathrm{Ind}_{C_1}^{G_1} \xi_1\overline{\xi_2}$, that is, the induced representation studied in the triple (G_1, C_1, ν) (cf. Chap. 5).

We now analyze, more closely, the $\mathrm{Ind}_{C_1}^{G_1} \xi_1\overline{\xi_2}$-component in (6.6). First of all, note that the representation space of $\mathrm{Ind}_{B_2}^{G_2} \chi_{\xi_1,\xi_2}$ is made up of all functions $F\colon G_2 \to \mathbb{C}$ such that

$$F\begin{pmatrix} x & y \\ 0 & z \end{pmatrix} = \overline{\xi_1(x)\xi_2(z)}F\begin{pmatrix} 1 & 0 \\ 0 & 1 \end{pmatrix} \tag{6.7}$$

and, if $w \neq 0$,

$$F\begin{pmatrix} x & y \\ w & z \end{pmatrix} = \overline{\xi_1(w)} \cdot \overline{\xi_2(x - yzw^{-1})}F\begin{pmatrix} xw^{-1} & 1 \\ 1 & 0 \end{pmatrix}, \tag{6.8}$$

where in the last identity we have used the Bruhat decomposition (cf. (3.32)):

$$\begin{pmatrix} x & y \\ w & z \end{pmatrix} = \begin{pmatrix} 1 & xw^{-1} \\ 0 & 1 \end{pmatrix}\begin{pmatrix} 0 & 1 \\ 1 & 0 \end{pmatrix}\begin{pmatrix} w & z \\ 0 & y - xzw^{-1} \end{pmatrix}.$$

On the other hand, the *natural* representation space of $\mathrm{Ind}_{C_1}^{G_1} \xi_1\overline{\xi_2}$ is made up of all functions $f\colon G_1 \to \mathbb{C}$ such that:

$$f\left(\begin{pmatrix} \alpha & \beta \\ \gamma & \delta \end{pmatrix}\begin{pmatrix} x_1 & \eta x_2 \\ x_2 & x_1 \end{pmatrix}\right) = \overline{\xi_1(x_1 + ix_2)} \cdot \overline{\xi_2(x_1 - ix_2)} \cdot f\begin{pmatrix} \alpha & \beta \\ \gamma & \delta \end{pmatrix}. \tag{6.9}$$

Proposition 6.2 *The* $\mathrm{Ind}_{C_1}^{G_1} \xi_1\overline{\xi_2}$*-component of* (6.6) *is made up of all* F *satisfying* (6.7) *and* (6.8), *supported in* $G_1 W B_2$ *(cf.* (6.2)*). Moreover, the map* $F \mapsto f$ *given by*

$$f\begin{pmatrix} \alpha & \beta \\ \gamma & \delta \end{pmatrix} = F\left(\begin{pmatrix} \alpha & \beta \\ \gamma & \delta \end{pmatrix} W\right) \tag{6.10}$$

for all $\begin{pmatrix} \alpha & \beta \\ \gamma & \delta \end{pmatrix} \in G_1$, *with* F *satisfying* (6.8), *is a linear bijection between the* $\mathrm{Ind}_{C_1}^{G_1} \xi_1\overline{\xi_2}$*-component of* (6.6) *and the natural realization of* $\mathrm{Ind}_{C_1}^{G_1} \xi_1\overline{\xi_2}$ *(see* (6.9)*). The inverse map is given by:* $f \mapsto F$, *with* f *satisfying* (6.9) *and* F, *supported in* $G_1 W B_2$, *is given by*

$$F\left(\begin{pmatrix} \alpha & \beta \\ \gamma & \delta \end{pmatrix} W \begin{pmatrix} x & y \\ 0 & z \end{pmatrix}\right) = \overline{\xi_1(x)} \cdot \overline{\xi_2(z)} \cdot f\begin{pmatrix} \alpha & \beta \\ \gamma & \delta \end{pmatrix} \tag{6.11}$$

for all $\begin{pmatrix} \alpha & \beta \\ \gamma & \delta \end{pmatrix} \in G_1$, $\begin{pmatrix} x & y \\ 0 & z \end{pmatrix} \in B_2$.

Proof From the proof of Theorem 6.1 and Mackey's lemma (see the formulation in [17, Section 11.5]) it follows that the $\mathrm{Ind}_{C_1}^{G_1}\xi_1\overline{\xi_2}$-component of $\mathrm{Res}_{G_1}^{G_2}\widehat{\chi}_{\xi_1,\xi_2}$ is made up of all functions in the representation space of $\widehat{\chi}_{\xi_1,\xi_2}$ that are supported in G_1WB_2. If F satisfies (6.8) and f is given by (6.10) then for all $\begin{pmatrix}\alpha & \beta\\ \gamma & \delta\end{pmatrix}\in G_1$, $\begin{pmatrix}x_1 & \eta x_2\\ x_2 & x_1\end{pmatrix}\in C_1$ we have:

$$\begin{aligned}
f\left(\begin{pmatrix}\alpha & \beta\\ \gamma & \delta\end{pmatrix}\begin{pmatrix}x_1 & \eta x_2\\ x_2 & x_1\end{pmatrix}\right) &= F\left(\begin{pmatrix}\alpha & \beta\\ \gamma & \delta\end{pmatrix}\begin{pmatrix}x_1 & \eta x_2\\ x_2 & x_1\end{pmatrix}W\right)\\
\text{(by (6.5))} &= F\left(\begin{pmatrix}\alpha & \beta\\ \gamma & \delta\end{pmatrix}W\begin{pmatrix}x_1+ix_2 & x_2\\ 0 & x_1-ix_2\end{pmatrix}\right)\\
\text{(by (6.8))} &= \overline{\xi_1(x_1+ix_2)}\cdot\overline{\xi_2(x_1-ix_2)}\cdot f\begin{pmatrix}\alpha & \beta\\ \gamma & \delta\end{pmatrix},
\end{aligned}$$

so that f satisfies (6.9). On the other hand, if f satisfies (6.9) and F is given by (6.11), then, for all $\begin{pmatrix}\alpha & \beta\\ \gamma & \delta\end{pmatrix}\in G_1$ and $\begin{pmatrix}x & y\\ 0 & z\end{pmatrix}\in B_2$, we have:

$$\begin{aligned}
F\left(\begin{pmatrix}\alpha & \beta\\ \gamma & \delta\end{pmatrix}W\begin{pmatrix}x & y\\ 0 & z\end{pmatrix}\right) &= \overline{\xi_1(x)}\cdot\overline{\xi_2(z)}\cdot f\begin{pmatrix}\alpha & \beta\\ \gamma & \delta\end{pmatrix}\\
&= \overline{\xi_1(x)}\cdot\overline{\xi_2(z)}\cdot F\left(\begin{pmatrix}\alpha & \beta\\ \gamma & \delta\end{pmatrix}W\right)
\end{aligned}$$

so that F satisfies (6.8). Clearly, the map $f\mapsto F$ is the inverse of the map $F\mapsto f$.

Remark 6.3 Let us take stock of the situation: in order to study $\mathrm{Ind}_{G_1}^{G_2}\rho_\nu$ we first have to describe the ρ_ν-component of $\mathrm{Res}_{G_1}^{G_2}\widehat{\chi}_{\xi_1,\xi_2}$ (cf. (6.1)). Then we use the machinery developed in Sect. 5.6 with $\nu_0=\xi_1\overline{\xi_2}$: the ρ_ν-component of $\mathrm{Res}_{G_1}^{G_2}\widehat{\chi}_{\xi_1,\xi_2}$ is clearly contained in the $\mathrm{Ind}_{C_1}^{G_1}\xi_1\overline{\xi_2}$-component in (6.6). As in the beginning of Sect. 2.2, we fix, once and for all, the vector $\delta_1\in V_{\rho_\nu}=L(\mathbb{F}_q^*)$ satisfying that (cf. (5.9))

$$\rho_\nu\begin{pmatrix}1 & y\\ 0 & 1\end{pmatrix}\delta_1=\chi(y)\delta_1. \tag{6.12}$$

In other words, δ_1 spans the χ-component of $\mathrm{Res}_{U_1}^{G_1}\rho_\nu$. Note that vector δ_1 is also used in the computation of the spherical function of the Gelfand–Graev representation (see Sect. 3.5 and, for an explicit computation, [17, Section 14.7]). Therefore, setting $\nu_0=\xi_1\overline{\xi_2}$, taking $f_0\in L(\mathbb{F}_q^*)$ as in Remark 5.5, we denote by $L\colon L(\mathbb{F}_q^*) \to \mathrm{Ind}_{C_1}^{G_1}\nu_0$ the associated intertwining operator (see [17, Equation (13.31)] which is just the particular case of Proposition 3.3 corresponding to the case $d_\theta = 1$ therein).

Lemma 6.3 *With the above notation, the function*

$$L\delta_1 \in \mathrm{Ind}_{C_1}^{G_1}\nu_0 \subseteq \mathrm{Res}_{G_1}^{G_2}\widehat{\chi}_{\xi_1,\xi_2}$$

is given by (modulo the identification $L\delta_1 = f_1 \mapsto F_1$ as in Proposition 6.2)

$$F_1\left(\begin{pmatrix}\alpha & \beta\\ 0 & 1\end{pmatrix}\begin{pmatrix}a & \eta b\\ b & a\end{pmatrix} w \begin{pmatrix}x & y\\ 0 & z\end{pmatrix}\right) = \frac{1}{\sqrt{q}}\overline{\xi_1(x(a+ib))}\cdot\overline{\xi_2(z(a-ib))} \cdot\chi(-y)\overline{f_0(x^{-1})} \tag{6.13}$$

for all $\begin{pmatrix}\alpha & \beta\\ 0 & 1\end{pmatrix}\begin{pmatrix}a & \eta b\\ b & a\end{pmatrix} \in G_1$ *(cf. Lemma 5.1) and* $\begin{pmatrix}x & y\\ 0 & z\end{pmatrix} \in B_2$.

Proof First of all, we compute $L\delta_1$: arguing as in the proof of Theorem 5.6, we have, for $g = \begin{pmatrix}x & y\\ 0 & 1\end{pmatrix}\begin{pmatrix}a & \eta b\\ b & a\end{pmatrix} \in G_1$,

$$\begin{aligned}[L\delta_1](g) &= \frac{1}{\sqrt{q}}\langle \delta_1, \rho_\nu(g) f_0\rangle_{L(\mathbb{F}_q^*)}\\ &= \frac{1}{\sqrt{q}}\sum_{z\in\mathbb{F}_q^*}\delta_1(z)\overline{\left[\rho_\nu\begin{pmatrix}x & y\\ 0 & 1\end{pmatrix}\rho_\nu\begin{pmatrix}a & \eta b\\ b & a\end{pmatrix} f_0\right](z)} \qquad (6.14)\\ &= \frac{1}{\sqrt{q}}\overline{\nu_0(a+ib)}\chi(-y)\overline{f_0(x^{-1})},\end{aligned}$$

where the first equality follows from [17, Equation (13.31)] (here $\sqrt{\frac{d_\theta}{|G/K|}} = \frac{1}{\sqrt{q}}$), and the last equality follows from (5.9), since $\rho_\nu\begin{pmatrix}a & \eta b\\ b & a\end{pmatrix} f_0 = \nu_0(a+ib) f_0$ (cf. (5.25)).

We now apply to the function $L\delta_1 \in \mathrm{Ind}_{C_1}^{G_1}\nu_0$ the map $f \mapsto F$ in Proposition 6.2, so that $L\delta_1 \mapsto F_1$ where

$$\begin{aligned}F_1\left(\begin{pmatrix}\alpha & \beta\\ 0 & 1\end{pmatrix}\begin{pmatrix}a & \eta b\\ b & a\end{pmatrix} w \begin{pmatrix}x & y\\ 0 & z\end{pmatrix}\right) &= \overline{\xi_1(x)}\cdot\overline{\xi_2(z)}[L\delta_1]\left(\begin{pmatrix}\alpha & \beta\\ 0 & 1\end{pmatrix}\begin{pmatrix}a & \eta b\\ b & a\end{pmatrix}\right)\\ &= \frac{1}{\sqrt{q}}\overline{\xi_1(x)}\cdot\overline{\xi_2(z)}\cdot\overline{\xi_1(a+ib)}\\ &\quad\cdot\overline{\xi_2(a-ib)}\cdot\chi(-y)\overline{f_0(x^{-1})},\end{aligned}$$

where the first equality follows from (6.11) and the second one follows from (6.14). This proves (6.13). □

Remark 6.4 With respect to the decomposition (6.4) we have

$$F_1\begin{pmatrix} a & b\\ c & d\end{pmatrix} = F_1\left(\begin{pmatrix} \alpha & \beta\\ 0 & 1\end{pmatrix} W \begin{pmatrix} c & d\\ 0 & (bc-ad)/(\alpha c)\end{pmatrix}\right)$$
$$= \frac{1}{\sqrt{q}}\overline{\xi_1(c)}\cdot\overline{\xi_2((bc-ad)/(\alpha c))}\cdot\chi(-d)\overline{f_0(c^{-1})}, \tag{6.15}$$

where $\begin{pmatrix} a & b\\ c & d\end{pmatrix} \in G_2$, $c \neq 0$, and $\alpha i + \beta = ac^{-1} \notin \mathbb{F}_q$ (note that here $a = 1$ and $b = 0$ in (6.13)).

Theorem 6.2 *Let $\xi_1, \xi_2 \in \widehat{\mathbb{F}^*_{q^2}}$ and suppose that they satisfy the hypotheses in Proposition 6.1. Then the spherical function of the triple (G_2, G_1, ρ_ν) associated with the parabolic representation $\widehat{\chi}_{\xi_1,\xi_2}$ and with the choice of the vector $\delta_1 \in L(\mathbb{F}_q^*)$ (cf. Remark 6.3) is given by*

$$\phi^{\xi_1,\xi_2}\begin{pmatrix} a & b\\ c & d\end{pmatrix} = \frac{1}{q(q+1)} \sum_{\substack{u=\alpha i+\beta\in\mathbb{F}_{q^2}\setminus\mathbb{F}_q:\\ cu+d\neq 0\\ \alpha_1 i+\beta_1=(au+b)/(cu+d)\notin\mathbb{F}_q}} \overline{\xi_1(cu+d)}\cdot\overline{\xi_2\left(\frac{\alpha(ad-bc)}{\alpha_1(cu+d)}\right)}$$
$$\cdot\,[\chi(-c)\cdot(-\nu(-(cu+d)^{-1}))$$
$$\cdot\sum_{\gamma\in\mathbb{F}_q}\overline{\xi_1(\gamma+i)}\cdot\overline{\xi_2(\gamma-i)}\chi(\gamma(cu+d+1))$$
$$\cdot\, j\left((cu+d)(\gamma^2-\eta)\right) + \delta_{(cu+d)^{-1}}(1)]$$

Proof We use (3.24), the decomposition for G_2 (cf. (3.32)), and the last expression in (1.24) to compute, for $g \in G_2$

$$\begin{aligned}\phi^{\xi_1,\xi_2}(g) &= \langle F_1, \lambda(g)F_1\rangle_{V_{\widehat{\chi}_{\xi_1,\xi_2}}}\\ &= \langle \lambda(g^{-1})F_1, F_1\rangle_{V_{\widehat{\chi}_{\xi_1,\xi_2}}}\\ &= \sum_{u\in U_2} F_1(guw)\overline{F_1(uw)} + F_1(g)\overline{F_1(1_{G_2})}\\ &= \sum_{u\in U_2} F_1(guw)\overline{F_1(uw)},\end{aligned} \tag{6.16}$$

where the last equality follows from the fact that F_1 is supported on $G_1WB_2 \not\ni 1_{G_2}$. Similarly, we need to determine when uw and guw belong to G_1WB_2. Now, if

$u = \begin{pmatrix} 1 & u \\ 0 & 1 \end{pmatrix}$ with $u \in \mathbb{F}_{q^2}$, by (6.3) we have that

$$uw = \begin{pmatrix} 1 & u \\ 0 & 1 \end{pmatrix}\begin{pmatrix} 0 & 1 \\ 1 & 0 \end{pmatrix} = \begin{pmatrix} u & 1 \\ 1 & 0 \end{pmatrix}$$

belongs to $G_1 W B_2$ if and only if $u \in \mathbb{F}_{q^2} \setminus \mathbb{F}_q$, and, if this is the case, its decomposition (cf. (6.4)) is given by

$$\begin{pmatrix} u & 1 \\ 1 & 0 \end{pmatrix} = \begin{pmatrix} \alpha & \beta \\ 0 & 1 \end{pmatrix}\begin{pmatrix} i & 1 \\ 1 & 0 \end{pmatrix}\begin{pmatrix} 1 & 0 \\ 0 & \alpha^{-1} \end{pmatrix}, \tag{6.17}$$

where $u = \alpha i + \beta$, with $\alpha, \beta \in \mathbb{F}_q$ and $\alpha \neq 0$.

If $g = \begin{pmatrix} a & b \\ c & d \end{pmatrix} \in G_2$, then

$$guw = \begin{pmatrix} a & b \\ c & d \end{pmatrix}\begin{pmatrix} u & 1 \\ 1 & 0 \end{pmatrix} = \begin{pmatrix} au+b & a \\ cu+d & c \end{pmatrix}$$

so that it belongs to $G_1 W B_2$ if an only if $cu + d \neq 0$ and $(au + b)/(cu + d) \notin \mathbb{F}_q$, and, if this is the case, its decomposition (cf. (6.4)) is given by

$$\begin{pmatrix} au+b & a \\ cu+d & c \end{pmatrix} = \begin{pmatrix} \alpha_1 & \beta_1 \\ 0 & 1 \end{pmatrix} W \begin{pmatrix} cu+d & c \\ 0 & (ad-bc)(\alpha_1(cu+d))^{-1} \end{pmatrix}, \tag{6.18}$$

where $\alpha_1 i + \beta_1 = (au + b)/(cu + d)$, with $\alpha_1, \beta_1 \in \mathbb{F}_q$ and $\alpha_1 \neq 0$.

Then, by virtue of (6.17), (6.18), (6.13), and (6.15), formula (6.16) becomes

$$\begin{aligned}\phi^{\xi_1,\xi_2}\begin{pmatrix} a & b \\ c & d \end{pmatrix} &= \frac{1}{q} \sum_{\substack{u \in \mathbb{F}_{q^2} \setminus \mathbb{F}_q: \\ cu+d \neq 0 \\ (au+b)/(cu+d) \notin \mathbb{F}_q}} \overline{\xi_1(cu+d)} \cdot \overline{\xi_2((ad-bc)(\alpha_1(cu+d))^{-1})} \\ &\quad \cdot \chi(-c)\overline{f_0((cu+d)^{-1})} \cdot \xi_2(\alpha^{-1}) f_0(1) \\ \text{(by (5.24))} &= \frac{1}{q} \sum_{\substack{u \in \mathbb{F}_{q^2} \setminus \mathbb{F}_q: \\ cu+d \neq 0 \\ (au+b)/(cu+d) \notin \mathbb{F}_q}} \overline{\xi_1(cu+d)} \cdot \overline{\xi_2(\alpha(ad-bc)(\alpha_1(cu+d))^{-1})} \\ &\quad \cdot \chi(-c) F_0((cu+d)^{-1}, 1).\end{aligned}$$

We then end the proof by invoking the explicit expression of $F_0(x, y)$ given by (5.22). □

6.2 Spherical Functions for $(\mathrm{GL}(2, \mathbb{F}_{q^2}), \mathrm{GL}(2, \mathbb{F}_q), \rho_\nu)$: the Cuspidal Case

We now examine the cuspidal representations in (6.1). We fix an indecomposable character $\mu \in \widehat{\mathbb{F}_{q^4}^*}$ such that $\mu^\sharp = \nu^\sharp$, where $\mu^\sharp = \mathrm{Res}_{\mathbb{F}_q^*}^{\mathbb{F}_{q^4}^*} \mu$ and $\nu^\sharp = \mathrm{Res}_{\mathbb{F}_q^*}^{\mathbb{F}_{q^2}^*} \nu$, as in Sect. 5.3.

We also define $\widetilde{\chi} \in \widehat{\mathbb{F}_{q^2}}$ by setting

$$\widetilde{\chi}(x + iy) = \chi(x) \text{ for all } x + iy \in \mathbb{F}_{q^2}, \tag{6.19}$$

where χ is the same nontrivial additive character of $\mathbb{F}_q$ fixed in Sect. 5.3 in order to define the generalized Kloosterman sum $j = j_{\chi,\nu}$. Then $\widetilde{\chi}$ is a nontrivial additive character of $\mathbb{F}_{q^2}$ and we can use it to define the corresponding Kloosterman sum $J = J_{\widetilde{\chi},\mu}$ over $\mathbb{F}_{q^2}^*$:

$$J(x) = \frac{1}{q^2} \sum_{\substack{w \in \mathbb{F}_{q^4}^*: \\ w\overline{w} = x}} \widetilde{\chi}(w + \overline{w})\mu(w)$$

for all $x \in \mathbb{F}_{q^2}^*$ (here, $\overline{w}$ indicates the conjugate of $w \in \mathbb{F}_{q^4}$: we regard $\mathbb{F}_{q^4}$ as a quadratic extension of $\mathbb{F}_{q^2}$). The function J is then used in (5.10) in order to define the cuspidal representation ρ_μ of G_2 (whose representation space is now $L(\mathbb{F}_{q^2}^*)$).

Our main goal is to study the restriction $\mathrm{Res}_{G_1}^{G_2} \rho_\mu$. We take a preliminary step which leads to some operators that simplify the calculations. More precisely, we start from the following fundamental fact: $\mathrm{Res}_{B_1}^{G_1} \rho_\nu$ is an irreducible B_1-representation (see [17, Section 14.6, in particular Theorem 14.6.9] for a more precise description of this representation. Observe that, in fact, it is given by (5.9)). Then we want to study the restriction $\mathrm{Res}_{B_1}^{G_2} \rho_\mu$.

Set $\Gamma = \{x + iy \in \mathbb{F}_{q^2}^* : x = 0\} \equiv \{iy : y \in \mathbb{F}_q^*\}$ and, for $\theta \in \mathbb{F}_q$, define $\Omega_\theta = \{x + iy \in \mathbb{F}_{q^2}^* : x \neq 0, y/x = \theta\} \equiv \{x + i\theta x : x \in \mathbb{F}_q^*\}$. Clearly,

$$\mathbb{F}_{q^2}^* = \Gamma \bigsqcup \left(\bigsqcup_{\theta \in \mathbb{F}_q} \Omega_\theta \right).$$

We also set $W = L(\Gamma)$ and, for $\theta \in \mathbb{F}_q$, define $V_\theta = L(\Omega_\theta)$.

Lemma 6.4

$$L(\mathbb{F}_{q^2}^*) = W \bigoplus \left(\bigoplus_{\theta \in \mathbb{F}_q} V_\theta \right)$$

is the decomposition of $L(\mathbb{F}_{q^2}^*)$ into $\mathrm{Res}_{B_1}^{G_2} \rho_\mu$-invariant subspaces. Moreover, each V_θ, $\theta \in \mathbb{F}_q$, is B_1-irreducible and isomorphic to $\mathrm{Res}_{B_1}^{G_1} \rho_\nu$, while $W = \bigoplus_{\substack{\psi_1, \psi_2 \in \widehat{\mathbb{F}_q^*}: \\ \psi_1\psi_2 = \mu^\sharp}} V_{\chi_{\psi_1,\psi_2}}$ is the sum of one-dimensional B_1-representations (cf. (5.3)). Finally, for each $\theta \in \mathbb{F}_q$, the operator

$$L_\theta \colon L(\mathbb{F}_q^*) \to L(\mathbb{F}_{q^2}^*)$$

defined by setting

$$[L_\theta f](x+iy) = \begin{cases} f((1-\theta^2\eta)x) & \text{if } x+iy \in \Omega_\theta \\ 0 & \text{otherwise} \end{cases} \tag{6.20}$$

for all $f \in L(\mathbb{F}_q^*)$ and $x + iy \in \mathbb{F}_{q^2}^*$, intertwines $\mathrm{Res}_{B_1}^{G_1} \rho_\nu$ with $\mathrm{Res}_{B_1}^{G_2} \rho_\mu$ (and, clearly, the image of L_θ is precisely V_θ).

Proof First of all, from (5.9) we deduce that for $\begin{pmatrix} \alpha & \beta \\ 0 & \delta \end{pmatrix} \in B_1$, $f \in L(\mathbb{F}_{q^2}^*)$, and $x + iy \in \mathbb{F}_{q^2}^*$

$$\begin{aligned} \left[\rho_\mu \begin{pmatrix} \alpha & \beta \\ 0 & \delta \end{pmatrix} f\right](x+iy) &= \mu(\delta)\widetilde{\chi}(\delta^{-1}\beta(x+iy)^{-1}) f(\delta\alpha^{-1}(x+iy)) \\ (\text{since } \mu^\sharp = \nu^\sharp) &= \nu(\delta)\widetilde{\chi}(\delta^{-1}\beta(x+iy)^{-1}) f(\delta\alpha^{-1}(x+iy)). \end{aligned} \tag{6.21}$$

Now, if we choose $\psi \in \widehat{\mathbb{F}_q^*}$ and we define $f \in W$ by setting

$$f(iy) = \psi(y) \quad \text{for all } y \in \mathbb{F}_q^*$$

then

$$\begin{aligned} \left[\rho_\mu \begin{pmatrix} \alpha & \beta \\ 0 & \delta \end{pmatrix} f\right](x+iy) &= \mu(\delta)\widetilde{\chi}(\delta^{-1}\beta(x+iy)^{-1}) f(\delta\alpha^{-1}(x+iy)) \\ (\text{since } \widetilde{\chi}(iy) = 1) &= \begin{cases} 0 & \text{if } x \neq 0 \\ \mu(\delta)\psi(\delta\alpha^{-1}y) & \text{otherwise} \end{cases} \\ &= \begin{cases} 0 & \text{if } x \neq 0 \\ \mu(\delta)\psi(\delta)\psi(\alpha^{-1})\psi(y) & \text{otherwise} \end{cases} \\ (\text{setting } \psi_1 = \overline{\psi} \text{ and } \psi_2 = \mu^\sharp\psi) &= \begin{cases} 0 & \text{if } x \neq 0 \\ \psi_1(\alpha)\psi_2(\delta)\psi(y) & \text{otherwise.} \end{cases} \end{aligned}$$

In other words, $\rho_\mu\begin{pmatrix}\alpha & \beta\\ 0 & \delta\end{pmatrix} f=\psi_1(\alpha)\psi_2(\delta)f$, and this proves the statement relative to W. On the other hand, invariance of V_θ, $\theta\in\mathbb{F}_q$, follows from (6.21) since if $f\in V_\theta$ then the function $f_1(x+iy)=f(\delta\alpha^{-1}(x+iy))$ still belongs to V_θ. Indeed $f_1=0$ unless $x\neq 0$, and $y/x\equiv(\delta\alpha^{-1}y)/(\delta\alpha^{-1}x)=\theta$.

It remains to show that L_θ intertwines $\mathrm{Res}^{G_1}_{B_1}\rho_\nu$ with the restriction of $\mathrm{Res}^{G_2}_{B_1}\rho_\mu$ on V_θ. We shall use the elementary identity

$$x+i\theta x=x(1-i\theta)^{-1}(1-\eta\theta^2). \tag{6.22}$$

For all $\begin{pmatrix}\alpha & \beta\\ 0 & \delta\end{pmatrix}\in B_1$, $f\in L(\mathbb{F}_q^*)$, $x\in\mathbb{F}_q^*$, and $\theta\in\mathbb{F}_q$ we have, using (6.21),

$$\left[\rho_\mu\begin{pmatrix}\alpha & \beta\\ 0 & \delta\end{pmatrix} L_\theta f\right](x+i\theta x)=\nu(\delta)\widetilde{\chi}(\delta^{-1}\beta(x+i\theta x)^{-1})[L_\theta f](\delta\alpha^{-1}(x+i\theta x))$$

$$\text{(by (6.20), (6.22), and (6.19))}\ =\nu(\delta)\chi(\delta^{-1}\beta x^{-1}(1-\eta\theta^2)^{-1})f(\delta\alpha^{-1}x(1-\eta\theta^2)).$$

On the other hand,

$$\left[L_\theta\rho_\nu\begin{pmatrix}\alpha & \beta\\ 0 & \delta\end{pmatrix} f\right](x+i\theta x)=\left[\rho_\nu\begin{pmatrix}\alpha & \beta\\ 0 & \delta\end{pmatrix} f\right]\left(x(1-\eta\theta^2)\right)$$

$$=\nu(\delta)\chi(\delta^{-1}\beta x^{-1}(1-\eta\theta^2)^{-1})f(\delta\alpha^{-1}x(1-\eta\theta^2)).$$

That is, $\rho_\mu\begin{pmatrix}\alpha & \beta\\ 0 & \delta\end{pmatrix} L_\theta f=L_\theta\rho_\nu\begin{pmatrix}\alpha & \beta\\ 0 & \delta\end{pmatrix} f$, and this ends the proof. □

Corollary 6.1 *The operators L_θ, $\theta\in\mathbb{F}_q$, form a basis for* $\mathrm{Hom}_{B_1}(\mathrm{Res}^{G_1}_{B_1}\rho_\nu,\mathrm{Res}^{G_2}_{B_1}\rho_\mu)$. □

In the following lemma we collect three basic, elementary, but quite useful identities.

Lemma 6.5 *For all* $\begin{pmatrix}x & y\\ 0 & z\end{pmatrix}\in B_1$, $u\in\mathbb{F}_q^*$, *and* $\theta\in\mathbb{F}_q$, *we have*

$$L_\theta\delta_u=\delta_{u(1-i\theta)^{-1}} \tag{6.23}$$

$$\rho_\nu\begin{pmatrix}x & y\\ 0 & z\end{pmatrix}\delta_u=\nu(z)\chi(yz^{-1}u^{-1})\delta_{uxz^{-1}} \tag{6.24}$$

$$L_\theta\rho_\nu\begin{pmatrix}x & y\\ 0 & z\end{pmatrix}\delta_u=\nu(z)\chi(yz^{-1}u^{-1})\delta_{uxz^{-1}(1-i\theta)^{-1}} \tag{6.25}$$

Proof For instance, $[L_\theta \delta_u](x + i\theta x) = \delta_u(x(1-\theta^2\eta))$ and $x(1-\theta^2\eta) = u$ yields

$$x + i\theta x = \frac{u}{1-\theta^2\eta}(1+i\theta) = u(1-i\theta)^{-1}.$$

□

We now introduce three operators. First of all, we define $Q_1 \colon L(\mathbb{F}^*_{q^2}) \to L(\mathbb{F}^*_{q^2})$ by setting

$$Q_1 f = \sum_{\theta \in \mathbb{F}_q} \langle f, L_\theta \delta_1 \rangle_{L(\mathbb{F}^*_{q^2})} L_\theta \delta_1 \tag{6.26}$$

for all $f \in L(\mathbb{F}^*_{q^2})$, where $\delta_1 \in L(\mathbb{F}^*_q)$, as in (6.12). Clearly, Q_1 is an orthogonal projection. Then we set

$$P = \frac{q-1}{|G_1|} \sum_{g \in G_1} \overline{\chi^\nu(g)} \rho_\mu(g), \tag{6.27}$$

where χ^ν is the character of ρ_ν. By (1.18) this is the projection of $L(\mathbb{F}^*_{q^2})$ onto its subspace, in the restriction $\mathrm{Res}^{G_2}_{G_1} \rho_\mu$, isomorphic to ρ_ν.

Finally, we assume that $L \colon L(\mathbb{F}^*_q) \to L(\mathbb{F}^*_{q^2})$ is an *isometric* immersion such that

$$L\rho_\nu(g) = \rho_\mu(g) L \quad \text{for all } g \in G_1 \tag{6.28}$$

(note that L coincides with the operator L_σ used in Sect. 3.3). Actually, we are only able to find an explicit formula for an operator L which is *not* isometric (see Sect. 6.3). However, an explicit formula for L (isometric or not) is not necessary in order to compute the spherical functions.

We just note two basic facts: on the one hand, since L spans $\mathrm{Hom}_{G_1}(\rho_\nu, \mathrm{Res}^{G_2}_{G_1} \rho_\mu)$ it also belongs to $\mathrm{Hom}_{B_1}(\mathrm{Res}^{G_1}_{B_1} \rho_\nu, \mathrm{Res}^{G_2}_{B_1} \rho_\mu)$ and therefore, by Corollary 6.1, there exist coefficients $\varphi(\theta)$, $\theta \in \mathbb{F}_q$, such that

$$L = \sum_{\theta \in \mathbb{F}_q} \varphi(\theta) L_\theta. \tag{6.29}$$

On the other hand, $\{\delta_x : x \in \mathbb{F}^*_q\}$ is an orthonormal basis of $L(\mathbb{F}^*_q)$ and therefore $\{L\delta_x : x \in \mathbb{F}^*_q\}$ is an orthonormal basis for the subspace of $L(\mathbb{F}^*_{q^2})$ which is $\mathrm{Res}^{G_2}_{G_1} \rho_\mu$-isomorphic to V_{ρ_ν}, so that

$$Pf = \sum_{x \in \mathbb{F}^*_q} \langle f, L\delta_x \rangle_{L(\mathbb{F}^*_{q^2})} L\delta_x \tag{6.30}$$

for all $f \in L(\mathbb{F}^*_{q^2})$.

Theorem 6.3 *The operators Q_1 and P commute. Moreover, setting $P_1 = PQ_1$, we have*

$$P_1 f = \langle f, L\delta_1 \rangle_{L(\mathbb{F}^*_{q^2})} L\delta_1 \tag{6.31}$$

for all $f \in L(\mathbb{F}^*_{q^2})$.

Finally, setting

$$F_1((1-i\theta)^{-1}, (1-i\sigma)^{-1}) = [PL_\theta\delta_1]((1-i\sigma)^{-1}) = [P\delta_{(1-i\theta)^{-1}}]((1-i\sigma)^{-1}) \tag{6.32}$$

for all $\theta, \sigma \in \mathbb{F}_q$, we have, for $f \in L(\mathbb{F}^*_{q^2})$:

$$[P_1 f]((1-i\sigma)^{-1}) = \sum_{\theta \in \mathbb{F}_q} F_1((1-i\theta)^{-1}, (1-i\sigma)^{-1}) f((1-i\theta)^{-1}) \tag{6.33}$$

for all $\sigma \in \mathbb{F}_q$, while $[P_1 f](u+iv) = 0$ if $u+iv$ is not of the form $(1-i\sigma)^{-1}$ for some $\sigma \in \mathbb{F}_q$.

Proof First of all, we prove that

$$\langle L_\theta \delta_x, L\delta_z \rangle = 0 \text{ if } x \neq z. \tag{6.34}$$

Indeed, since L_θ, $\theta \in \mathbb{F}_q$, and L belong to $\mathrm{Hom}_{B_1}(\mathrm{Res}^{G_1}_{B_1}\rho_\nu, \mathrm{Res}^{G_2}_{B_1}\rho_\mu)$ (cf. Corollary 6.1 and (6.29)) and $U_1 \subseteq B_1$, we have, for all $x \in \mathbb{F}^*_q$

$$\rho_\mu \begin{pmatrix} 1 & y \\ 0 & 1 \end{pmatrix} L_\theta \delta_x = L_\theta \rho_\nu \begin{pmatrix} 1 & y \\ 0 & 1 \end{pmatrix} \delta_x = \chi(x^{-1}y) L_\theta \delta_x,$$

where the last equality follows from (6.24), and, similarly,

$$\rho_\mu \begin{pmatrix} 1 & y \\ 0 & 1 \end{pmatrix} L\delta_x = \chi(x^{-1}y) L\delta_x.$$

In other words, setting $\chi_x(y) = \chi(xy)$ (cf. [17, Proposition 7.1.1]), then $L_\theta\delta_x$ and $L\delta_x$ belong to the $(\chi_{x^{-1}})$-isotypic component of $\mathrm{Res}^{G_2}_{U_1}\rho_\mu$. Then, (6.34) just expresses the orthogonality between distinct U_1-isotypic components.

Note that we may also deduce that Q_1, defined by (6.26), is just the orthogonal projection onto the χ-isotypic component of $\mathrm{Res}^{G_2}_{U_1}\rho_\mu$. In particular

$$Q_1 L\delta_1 = L\delta_1. \tag{6.35}$$

Alternatively, it is easy to deduce (6.35) directly from (6.29) and (6.26).

Let $f \in L(\mathbb{F}_{q^2}^*)$. Then, from (6.26) and (6.30) we deduce that

$$\begin{aligned}
Q_1 P f &= \sum_{\theta \in \mathbb{F}_q} \sum_{x \in \mathbb{F}_q^*} \langle f, L\delta_x \rangle \cdot \langle L\delta_x, L_\theta \delta_1 \rangle L_\theta \delta_1 \\
\text{(by (6.34))} &= \langle f, L\delta_1 \rangle \sum_{\theta \in \mathbb{F}_q} \langle L\delta_1, L_\theta \delta_1 \rangle L_\theta \delta_1 \\
\text{(by (6.26))} &= \langle f, L\delta_1 \rangle Q_1 L\delta_1 \\
\text{(by (6.35))} &= \langle f, L\delta_1 \rangle L\delta_1.
\end{aligned}$$

Similarly,

$$\begin{aligned}
P Q_1 f &= \sum_{\theta \in \mathbb{F}_q} \sum_{x \in \mathbb{F}_q^*} \langle f, L_\theta \delta_1 \rangle \cdot \langle L_\theta \delta_1, L\delta_x \rangle L\delta_x \\
\text{(by (6.34))} &= \sum_{\theta \in \mathbb{F}_q} \langle f, L_\theta \delta_1 \rangle \cdot \langle L_\theta \delta_1, L\delta_1 \rangle L\delta_1 \\
&= \langle f, \sum_{\theta \in \mathbb{F}_q} \langle L\delta_1, L_\theta \delta_1 \rangle L_\theta \delta_1 \rangle L\delta_1 \\
\text{(by (6.26))} &= \langle f, Q_1 L\delta_1 \rangle L\delta_1 \\
\text{(by (6.35))} &= \langle f, L\delta_1 \rangle L\delta_1.
\end{aligned}$$

In conclusion, we have proved the equality $PQ_1 = Q_1 P$ and (6.31).

Moreover, as $P_1 = Q_1 P Q_1$, for $f \in L(\mathbb{F}_{q^2}^*)$ we have

$$\begin{aligned}
P_1 f &= Q_1 P Q_1 f \\
\text{(by (6.26))} &= \sum_{\sigma \in \mathbb{F}_q} \langle P Q_1 f, L_\sigma \delta_1 \rangle L_\sigma \delta_1 \\
\text{(again by (6.26))} &= \sum_{\theta, \sigma \in \mathbb{F}_q} \langle f, L_\theta \delta_1 \rangle \cdot \langle P L_\theta \delta_1, L_\sigma \delta_1 \rangle L_\sigma \delta_1 \\
\text{(by (6.23))} &= \sum_{\theta, \sigma \in \mathbb{F}_q} f((1 - i\theta)^{-1}) \cdot \langle P\delta_{(1-i\theta)^{-1}}, \delta_{(1-i\sigma)^{-1}} \rangle \delta_{(1-i\sigma)^{-1}}
\end{aligned}$$

and (6.33) follows as well. □

Corollary 6.2 *For $u, v \in \mathbb{F}_q$ one has $[L\delta_1](u + iv) = 0$ if $u + iv$ is not of the form $(1 - i\sigma)^{-1}$ for some $\sigma \in \mathbb{F}_q$.*

Proof This follows from the fact that for all $f \in L(\mathbb{F}^*_{q^2})$ one has $[P_1 f](u+iv) = 0$ if $u + iv$ is not of the form $(1 - i\sigma)^{-1}$ for some $\sigma \in \mathbb{F}_q$, and that, after taking $f = L\delta_1$, one has $[L\delta_1](u + iv) = [P_1 f](u + iv)$, by (6.31). □

Corollary 6.3 *For $\sigma, \theta \in \mathbb{F}_q$ one has (cf. (6.32))*

$$\overline{[L\delta_1]((1 - i\theta)^{-1})} \cdot [L\delta_1]((1 - i\sigma)^{-1}) = \acute{F}_1((1 - i\theta)^{-1}, (1 - i\sigma)^{-1}).$$

Proof This is an immediate consequence of (6.31) (with $f = L\delta_1$) and (6.33). □

Note that the expression in the above corollary is the analogue, in the present setting, of (5.24).

The following lemmas lead to a considerable simplification in the application of (6.27).

Lemma 6.6 *Let $\begin{pmatrix} a & b \\ c & d \end{pmatrix} \in G_1$ with $c \neq 0$. Then we have a unique factorization*

$$\begin{pmatrix} a & b \\ c & d \end{pmatrix} = \begin{pmatrix} \gamma & \eta \\ 1 & \gamma \end{pmatrix} \begin{pmatrix} x & y \\ 0 & z \end{pmatrix}$$

with $\gamma, y \in \mathbb{F}_q$ and $x, z \in \mathbb{F}_q^*$.

Proof The equality is equivalent to the system

$$\begin{cases} \gamma x = a \\ \gamma y + \eta z = b \\ x = c \\ y + \gamma z = d \end{cases}$$

which admits a unique solution, namely: $x = c$, $\gamma = ax^{-1}$, and y, z may be computed by means of the Cramer rule, taking into account that $\det \begin{pmatrix} \gamma & \eta \\ 1 & \gamma \end{pmatrix} = \gamma^2 - \eta \neq 0$, since η is not a square. □

Lemma 6.7 *Let $\begin{pmatrix} a & b \\ c & d \end{pmatrix} \in G_1$ with $c \neq 0$. Then we have*

$$\sum_{y \in \mathbb{F}_q} \overline{\chi^\nu\left(\begin{pmatrix} a & b \\ c & d \end{pmatrix} \begin{pmatrix} 1 & y \\ 0 & 1 \end{pmatrix}\right)} \rho_\nu \begin{pmatrix} 1 & y \\ 0 & 1 \end{pmatrix} \delta_1 = -q\nu(c(ad - bc)^{-1})$$

$$\cdot \chi(-ac^{-1} - dc^{-1}) j(c^{-2}(ad - bc))\delta_1.$$

Proof Using the orthonormal basis $\{\delta_u : u \in \mathbb{F}_q^*\}$ in $L(\mathbb{F}_q^*)$ to compute the character χ^ν, we find:

$$\sum_{y\in\mathbb{F}_q} \overline{\chi^\nu\left(\begin{pmatrix} a & b \\ c & d \end{pmatrix}\begin{pmatrix} 1 & y \\ 0 & 1 \end{pmatrix}\right)} \rho_\nu\begin{pmatrix} 1 & y \\ 0 & 1 \end{pmatrix}\delta_1 = \sum_{y\in\mathbb{F}_q}\sum_{u\in\mathbb{F}_q^*}\left\langle \delta_u, \rho_\nu\begin{pmatrix} a & b \\ c & d \end{pmatrix}\rho_\nu\begin{pmatrix} 1 & y \\ 0 & 1 \end{pmatrix}\delta_u\right\rangle \cdot \rho_\nu\begin{pmatrix} 1 & y \\ 0 & 1 \end{pmatrix}\delta_1$$

$$\text{(by (6.24))} = \sum_{u\in\mathbb{F}_q^*}\left[\sum_{y\in\mathbb{F}_q}\chi(y(1-u^{-1}))\right]\left\langle \delta_u, \rho_\nu\begin{pmatrix} a & b \\ c & d \end{pmatrix}\delta_u\right\rangle\delta_1$$

$$\text{(orthogonality relations infer } u=1) = q\left\langle \delta_1, \rho_\nu\begin{pmatrix} a & b \\ c & d \end{pmatrix}\delta_1\right\rangle\delta_1$$

$$\text{(setting } x=y=1 \text{ in (5.10))} = -q\nu(-c^{-1})\chi(-ac^{-1}-dc^{-1}) \cdot \overline{j(c^{-2}(ad-bc))}\delta_1$$

and, by means of the identity (5.8), one immediately obtains the desired expression in the statement. □

Remark 6.5 Note that the coefficient of δ_1 in the formula of the above lemma is q-times the spherical function of the Gelfand–Graev representation (see Theorem 3.9) associated with the cuspidal representation ρ_ν; see [17, Proposition 14.7.9.(iii)].

We are now in a position to face up to the most difficult computation of this chapter. Recall that j (resp. J) is the generalized Kloosterman sum over $\mathbb{F}_q^*$ (resp. $\mathbb{F}_{q^2}^*$).

Theorem 6.4 *The function F_1 in Theorem 6.3 has the following expression:*

$$F_1((1-i\theta)^{-1}, (1-i\sigma)^{-1}) = \frac{1}{q+1}\delta_{\theta,\sigma} + \frac{q}{q+1}\sum_{t\in\mathbb{F}_q^*}\mu(-t^{-1}(1-i\theta)^{-1}) \cdot j(t)J(t(1-i\sigma)(1-i\theta)),$$

for all $\theta, \sigma \in \mathbb{F}_q$.

Proof We make use of the following facts that lead to a considerable simplification. First of all, we have (cf. Corollary 6.1)

$$\rho_\mu\begin{pmatrix} x & y \\ 0 & z \end{pmatrix} L_\theta = L_\theta\rho_\nu\begin{pmatrix} x & y \\ 0 & z \end{pmatrix} \tag{6.36}$$

for all $\theta \in \mathbb{F}_q$, $x, z \in \mathbb{F}_q^*$, and $y \in \mathbb{F}_q$.

Moreover, since $\mathrm{Res}^{G_1}_{B_1}\rho_\nu$ is irreducible (cf. [17, Section 14.6]), then the orthogonal projection formula (1.18) gives:

$$\frac{q-1}{|B_1|}\sum_{\left(\begin{smallmatrix} x & y\\ 0 & z\end{smallmatrix}\right)\in B_1}\overline{\chi^\nu\begin{pmatrix} x & y\\ 0 & z\end{pmatrix}}\rho_\nu\begin{pmatrix} x & y\\ 0 & z\end{pmatrix}=I_{L(\mathbb{F}_q^*)},\tag{6.37}$$

where $I_{L(\mathbb{F}_q^*)}$ is the identity operator on $L(\mathbb{F}_q^*)$.

Then, starting from (6.32), we get

$$\begin{aligned}
F_1((1-i\theta)^{-1},(1-i\sigma)^{-1})&=[PL_\theta\delta_1]((1-i\sigma)^{-1})\\
\text{(by (6.27))}&=\frac{1}{q(q^2-1)}\sum_{g\in G_1}\overline{\chi^\nu(g)}[\rho_\mu(g)L_\theta\delta_1]((1-i\sigma)^{-1})\\
\text{(by (3.32), Lemma 6.6)}&=\frac{1}{q(q^2-1)}\sum_{\left(\begin{smallmatrix} x & y\\ 0 & z\end{smallmatrix}\right)\in B_1}\overline{\chi^\nu\begin{pmatrix} x & y\\ 0 & z\end{pmatrix}}\\
&\quad\cdot\left[\rho_\mu\begin{pmatrix} x & y\\ 0 & z\end{pmatrix}L_\theta\delta_1\right](1-i\sigma)^{-1})\\
&\quad+\frac{1}{q(q^2-1)}\sum_{\gamma\in\mathbb{F}_q}\sum_{\substack{x,z\in\mathbb{F}_q^*\\ y\in\mathbb{F}_q}}\overline{\chi^\nu\left(\begin{pmatrix} \gamma & \eta\\ 1 & \gamma\end{pmatrix}\begin{pmatrix} x & y\\ 0 & z\end{pmatrix}\right)}\\
&\quad\cdot[\rho_\mu\begin{pmatrix} \gamma & \eta\\ 1 & \gamma\end{pmatrix}\rho_\mu\begin{pmatrix} x & y\\ 0 & z\end{pmatrix}L_\theta\delta_1](1-i\sigma)^{-1})\\
\text{(by (6.36) and (6.37))}&=\frac{1}{q+1}\left[L_\theta I_{L(\mathbb{F}_q^*)}\delta_1\right]((1-i\sigma)^{-1})\\
&\quad+\frac{1}{q(q^2-1)}\sum_{\gamma\in\mathbb{F}_q}\sum_{\substack{x,z\in\mathbb{F}_q^*\\ y\in\mathbb{F}_q}}\overline{\chi^\nu\left(\begin{pmatrix} \gamma & \eta\\ 1 & \gamma\end{pmatrix}\begin{pmatrix} x & y\\ 0 & z\end{pmatrix}\right)}\\
&\quad\cdot\left[\rho_\mu\begin{pmatrix} \gamma & \eta\\ 1 & \gamma\end{pmatrix}L_\theta\rho_\nu\begin{pmatrix} x & y\\ 0 & z\end{pmatrix}\delta_1\right](1-i\sigma)^{-1})\\
\text{(by (6.23))}&=\frac{1}{q+1}\delta_{\theta,\sigma}+\frac{1}{q(q^2-1)}\sum_{\gamma\in\mathbb{F}_q}\sum_{x,z\in\mathbb{F}_q^*}\left[\rho_\mu\begin{pmatrix} \gamma & \eta\\ 1 & \gamma\end{pmatrix}L_\theta\rho_\nu\begin{pmatrix} x & 0\\ 0 & z\end{pmatrix}\right.\\
&\quad\cdot\sum_{y\in\mathbb{F}_q}\overline{\chi^\nu\left(\begin{pmatrix} \gamma & \eta\\ 1 & \gamma\end{pmatrix}\begin{pmatrix} x & 0\\ 0 & z\end{pmatrix}\begin{pmatrix} 1 & x^{-1}y\\ 0 & 1\end{pmatrix}\right)}\\
&\quad\cdot\left.\rho_\nu\begin{pmatrix} 1 & x^{-1}y\\ 0 & 1\end{pmatrix}\delta_1\right]((1-i\sigma)^{-1})
\end{aligned}$$

$$\begin{aligned}
\text{(by Lemma 6.7)} &= \frac{1}{q+1}\delta_{\theta,\sigma} - \frac{1}{(q^2-1)}\sum_{\gamma\in\mathbb{F}_q}\sum_{x,z\in\mathbb{F}_q^*}\nu(z^{-1}(\gamma^2-\eta)^{-1}) \\
&\quad\cdot\chi(-\gamma(1+zx^{-1}))j(x^{-1}z(\gamma^2-\eta))[\rho_\mu\begin{pmatrix}\gamma & \eta\\ 1 & \gamma\end{pmatrix} \\
&\quad\cdot L_\theta\rho_\nu\begin{pmatrix}x & 0\\ 0 & z\end{pmatrix}\delta_1]((1-i\sigma)^{-1}) \\
\text{(by (6.25))} &= \frac{1}{q+1}\delta_{\theta,\sigma} - \frac{1}{(q^2-1)}\sum_{\gamma\in\mathbb{F}_q}\sum_{x,z\in\mathbb{F}_q^*}\nu((\gamma^2-\eta)^{-1}) \\
&\quad\cdot\chi(-\gamma(1+zx^{-1}))j(x^{-1}z(\gamma^2-\eta)) \\
&\quad\cdot[\rho_\mu\begin{pmatrix}\gamma & \eta\\ 1 & \gamma\end{pmatrix}\delta_{xz^{-1}(1-i\theta)^{-1}}]((1-i\sigma)^{-1})
\end{aligned}$$

Actually, only the expression $x^{-1}z$ appears and therefore the sum over x may be omitted and a factor $q-1$ must be added.

$$\begin{aligned}
\text{(by (5.10))} &= \frac{1}{q+1}\delta_{\theta,\sigma} + \frac{1}{q+1}\sum_{\gamma\in\mathbb{F}_q}\sum_{z\in\mathbb{F}_q^*}\nu((\gamma^2-\eta)^{-1}) \\
&\quad\cdot\chi(-\gamma(1+z))j(z(\gamma^2-\eta))\sum_{v+iw\in\mathbb{F}_{q^2}^*}[\mu(-(v+iw)) \\
&\quad\cdot\widetilde{\chi}(\gamma(1-i\sigma)+\gamma(v+iw)^{-1}) \\
&\quad\cdot J((1-i\sigma)(v+iw)^{-1}(\gamma^2-\eta)) \\
&\quad\cdot\delta_{z^{-1}(1-i\theta)^{-1}}(v+iw)] \\
&= \frac{1}{q+1}\delta_{\theta,\sigma} + \frac{1}{q+1}\sum_{\gamma\in\mathbb{F}_q}\sum_{z\in\mathbb{F}_q^*}\nu((\gamma^2-\eta)^{-1})\chi(-\gamma(1+z)) \\
&\quad\cdot j(z(\gamma^2-\eta))[\mu(-z^{-1}(1-i\theta)^{-1})\chi(\gamma+\gamma z) \\
&\quad\cdot J((1-i\sigma)(1-i\theta)z(\gamma^2-\eta))],
\end{aligned}$$

where in the last equality we have used the fact that $v+iw=z^{-1}(1-i\theta)^{-1}$ and the equality $\widetilde{\chi}(\alpha+i\beta)=\chi(\alpha)$ which follows from (6.19). The two χ factors simplify and, introducing the new variable $t=z(\gamma^2-\eta)$ in place of z, also the variable γ disappears. Moreover, the sum over γ yields a factor q and, recalling that $\mu^\sharp=\nu^\sharp$, one gets the expression in the statement. □

Arguing as in the proof of Theorem 5.6, we use the projection formula (6.33), with the explicit expression of F_1 given in Theorem 6.4, to compute the spherical function associated with the cuspidal representation ρ_μ. We need an elementary, preliminary result.

Proposition 6.3 *Each number $\alpha + i\beta \in \mathbb{F}_{q^2}$ with $\beta \neq 0$ may be uniquely represented in the form $\alpha + i\beta = \frac{1-i\theta}{1-i\sigma}$, with $\theta, \sigma \in \mathbb{F}_q$. Moreover, 1 may be represented in q different ways (namely, setting $\theta = \sigma \in \mathbb{F}_q$), while $\alpha \in \mathbb{F}_q$, $\alpha \neq 1$, is not representable in this way.*

Proof The equation $\alpha + i\beta = \frac{1-i\theta}{1-i\sigma}$ leads to the system

$$\begin{cases} \alpha - \beta\sigma\eta &= 1 \\ \beta - \sigma\alpha &= -\theta \end{cases}$$

which, for $\beta \neq 0$, has the unique solution

$$\sigma = \frac{\alpha - 1}{\beta\eta}, \qquad \theta = \sigma\alpha - \beta.$$

□

Following the preceding proposition, we set $\mathfrak{S}_0 = \{\alpha + i\beta \in \mathbb{F}_q : \beta \neq 0\}$.

Theorem 6.5 *The spherical function ϕ^μ of the multiplicity-free triple (G_2, G_1, ρ_ν) associated with the cuspidal representation ρ_μ (where $\mu \in \widehat{\mathbb{F}^*_{q^4}}$ is indecomposable over $\mathbb{F}^*_{q^2}$ and $\mu^\sharp = \nu^\sharp$) has the following expression: for $\begin{pmatrix} a & b \\ c & d \end{pmatrix} \in G_2$,*

$$\phi^\mu \begin{pmatrix} a & b \\ 0 & d \end{pmatrix} = 0 \qquad \text{if } a \neq d \text{ and } da^{-1} \notin \mathfrak{S}_0; \tag{6.38}$$

$$\phi^\mu \begin{pmatrix} a & b \\ 0 & d \end{pmatrix} = \mu(d^{-1})\widetilde{\chi}(-d^{-1}b(1 - i\theta))F_1((1 - i\sigma)^{-1}, (1 - i\theta)^{-1}) \tag{6.39}$$

if $a \neq d$, $da^{-1} \in \mathfrak{S}_0$, and $da^{-1} = \frac{1-i\theta}{1-i\sigma}$;

$$\phi^\mu \begin{pmatrix} a & b \\ 0 & a \end{pmatrix} = \sum_{\theta \in \mathbb{F}_q} \mu(d^{-1})\widetilde{\chi}(-a^{-1}b(1 - i\theta))F_1((1 - i\theta)^{-1}, (1 - i\theta)^{-1}) \tag{6.40}$$

if $d = a$;

$$\begin{aligned} \phi^\mu \begin{pmatrix} a & b \\ c & d \end{pmatrix} = - \sum_{\theta, \sigma \in \mathbb{F}_q} & \mu(c(1 - i\sigma)^{-1}(ad - bc)^{-1})) \\ & \cdot \widetilde{\chi}(-ac^{-1}(1 - i\sigma) - dc^{-1}(1 - i\theta)) \\ & \cdot J(c^{-2}(1 - i\sigma)(1 - i\theta)(ad - bc))F_1((1 - i\theta)^{-1}, (1 - i\sigma)^{-1}) \end{aligned} \tag{6.41}$$

if $c \neq 0$.

Proof From (3.24), with $L\delta_1$ in place of w^σ, we get the general expression for $g \in G_2$:

$$\phi^\mu(g) = \langle L\delta_1, \rho_\mu(g)L\delta_1\rangle_{L(\mathbb{F}_{q^2}^*)} = \sum_{\theta\in\mathbb{F}_q} [L\delta_1]((1-i\theta)^{-1})\overline{[\rho_\mu(g)L\delta_1]((1-i\theta)^{-1})}, \tag{6.42}$$

where the last equality follows from Corollary 6.2. For $g = \begin{pmatrix} a & b \\ 0 & d \end{pmatrix} \in B_2$, applying (5.9) to (6.42) we have

$$\begin{aligned}
\phi^\mu(g) &= \sum_{\theta\in\mathbb{F}_q} [L\delta_1]((1-i\theta)^{-1})\mu(d^{-1})\widetilde{\chi}(-d^{-1}b(1-i\theta)) \\
&\quad \cdot \overline{[L\delta_1](da^{-1}(1-i\theta)^{-1})} \\
\text{(by Corollary 6.3)} &= \sum_{\theta\in\mathbb{F}_q} \mu(d^{-1})\widetilde{\chi}(-d^{-1}b(1-i\theta)) \\
&\quad \cdot F_1(da^{-1}(1-i\theta)^{-1}, (1-i\theta)^{-1}).
\end{aligned}$$

We now impose the condition in Corollary 6.2: $da^{-1}(1-i\theta)^{-1}$ must be of the form $(1-i\sigma)^{-1}$ for some $\sigma \in \mathbb{F}_q$. If $a = d$ then $\sigma = \theta$ and we get (6.40).

By Proposition 6.3, if $a \neq d$ then $da^{-1}(1-i\theta)^{-1} = (1-i\sigma)^{-1}$ for some $\sigma \in \mathbb{F}_q$ if and only if $da^{-1} \in \mathfrak{S}_0$ and we deduce (6.38) and (6.39).

Finally we examine the case $g = \begin{pmatrix} a & b \\ c & d \end{pmatrix} \in G_2$ with $c \neq 0$. Now by (5.10) applied to (6.42) we get

$$\begin{aligned}
\phi^\mu(g) &= -\sum_{\sigma\in\mathbb{F}_q} [L\delta_1]((1-i\sigma)^{-1}) \sum_{\theta\in\mathbb{F}_q} \overline{\mu(-(1-i\theta)^{-1}c)} \\
&\quad \cdot \overline{\widetilde{\chi}(ac^{-1}(1-i\sigma) + dc^{-1}(1-i\theta))} \\
&\quad \cdot \overline{J(c^{-2}(1-i\sigma)(1-i\theta)(ad-bc))} \\
&\quad \cdot \overline{[L\delta_1]((1-i\theta)^{-1})} \\
\text{(by (5.8) and Corollary 6.3)} &= -\sum_{\sigma,\theta\in\mathbb{F}_q} \mu(c(ad-bc)^{-1}(1-i\sigma)^{-1}) \\
&\quad \cdot \widetilde{\chi}(-ac^{-1}(1-i\sigma) - dc^{-1}(1-i\theta)) \\
&\quad \cdot J(c^{-2}(1-i\sigma)(1-i\theta)(ad-bc)) \\
&\quad \cdot F_1((1-i\theta)^{-1}, (1-i\sigma)^{-1}).
\end{aligned}$$

□

Remark 6.6 By setting $\theta = 0$ in F_1 (cf. Theorem 6.4) we get the vector

$$f_1(\sigma) = \frac{1}{q+1}\delta_{0,\sigma} + \frac{q}{q+1}\sum_{t\in\mathbb{F}_q^*}\mu(-t^{-1})j(t)J(t(1-i\sigma)) \tag{6.43}$$

which, by virtue of Corollary 6.3, is a non-normalized multiple of $L\delta_1$. Actually, it is not easy to compute the norm of f_1: we discuss this problem in the next section.

6.3 A Non-normalized $\widetilde{L} \in \mathrm{Hom}_{G_1}(\rho_\nu, \mathrm{Res}^{G_2}_{G_1}\rho_\mu)$

In this section, we want to describe an alternative approach to (6.43) by deriving a non-normalized multiple $\widetilde{L}$ of L in (6.28). This is also a way to revisit the results and the calculations in the preceding section. The idea is simple: we set

$$\widetilde{L} = \frac{1}{|G_1|}\sum_{g\in G_1}\rho_\mu(g)L_0\rho_\nu(g^{-1}) \tag{6.44}$$

where L_0 is as in (6.20). Then, for all $h \in G_1$,

$$\begin{aligned}\rho_\mu(h)\widetilde{L} &= \frac{1}{|G_1|}\sum_{g\in G_1}\rho_\mu(hg)L_0\rho_\nu(g^{-1})\\ (\text{setting } r = hg) \quad &= \frac{1}{|G_1|}\sum_{r\in G_1}\rho_\mu(r)L_0\rho_\nu(r^{-1})\rho_\nu(h)\\ &= \widetilde{L}\rho_\nu(h),\end{aligned}$$

that is, $\widetilde{L} \in \mathrm{Hom}_{G_1}(\rho_\nu, \mathrm{Res}^{G_2}_{G_1}\rho_\mu)$. The operator L_0 may be replaced by any linear operator $T\colon L(\mathbb{F}_q^*) \to L(\mathbb{F}_{q^2}^*)$ (but checking that, eventually, $\widetilde{L} \neq 0$), for instance by taking $T = L_\theta$, with $\theta \in \mathbb{F}_q$. The choice of L_0 greatly simplifies the calculation and the final formula. We split the explicit computation of (6.44) in several preliminary results.

Lemma 6.8 *Let $T\colon L(\mathbb{F}_q^*) \to L(\mathbb{F}_{q^2}^*)$ be a linear operator. Suppose that its (matrix) kernel is $(a(u, v))_{u\in\mathbb{F}_{q^2}^*, v\in\mathbb{F}_q^*}$, so that T is expressed by setting*

$$[Tf](x+iy) = \sum_{v\in\mathbb{F}_q^*} a(x+iy, v)f(v)$$

for all $f \in L(\mathbb{F}_q^*)$ *and* $x + iy \in \mathbb{F}_{q^2}^*$. *Then*

$$\left[\sum_{u\in\mathbb{F}_q} \rho_\mu \begin{pmatrix} 1 & u \\ 0 & 1 \end{pmatrix} T\rho_\nu \begin{pmatrix} 1 & -u \\ 0 & 1 \end{pmatrix} f\right](x+iy) = \begin{cases} 0 & \text{if } x = 0 \\ qa(x+iy, \frac{x^2-\eta y^2}{x}) f(\frac{x^2-\eta y^2}{x}) & \text{if } x \neq 0. \end{cases} \tag{6.45}$$

Proof By (5.9)

$$\begin{aligned}\left[\sum_{u\in\mathbb{F}_q} \rho_\mu \begin{pmatrix} 1 & u \\ 0 & 1 \end{pmatrix} T\rho_\nu \begin{pmatrix} 1 & -u \\ 0 & 1 \end{pmatrix} f\right](x+iy) &= \sum_{u\in\mathbb{F}_q} \widetilde{\chi}((x+iy)^{-1}u) \\ &\quad \cdot \left[T\rho_\nu \begin{pmatrix} 1 & -u \\ 0 & 1 \end{pmatrix} f\right](x+iy) \\ &= \sum_{u\in\mathbb{F}_q} \sum_{v\in\mathbb{F}_q^*} \widetilde{\chi}((x+iy)^{-1}u) a(x+iy, v) \\ &\quad \cdot \left[\rho_\nu \begin{pmatrix} 1 & -u \\ 0 & 1 \end{pmatrix} f\right](v) \\ &= \sum_{u\in\mathbb{F}_q} \sum_{v\in\mathbb{F}_q^*} \widetilde{\chi}(u[(x+iy)^{-1} - v^{-1}]) \\ &\quad \cdot a(x+iy, v) f(v).\end{aligned}$$

But from

$$\frac{1}{x+iy} = \frac{x}{x^2-\eta y^2} - i\frac{y}{x^2-\eta y^2},$$

definition (6.19), and the orthogonal relations in $\widehat{\mathbb{F}_q}$ it follows that, for $x \neq 0$,

$$\sum_{u\in\mathbb{F}_q} \widetilde{\chi}(u[(x+iy)^{-1} - v^{-1}]) = \sum_{u\in\mathbb{F}_q} \chi\left(u\left(\frac{x}{x^2-\eta y^2} - \frac{1}{v}\right)\right) = q\delta_{v, \frac{x}{x^2-\eta y^2}},$$

while this sum, for $x = 0$, is equal to $\sum_{u\in\mathbb{F}_q} \chi(-\frac{u}{v}) = 0$. Then formula (6.45) follows immediately. □

Recall that $w = \begin{pmatrix} 0 & 1 \\ 1 & 0 \end{pmatrix}$ (see the Bruhat decomposition (3.32)).

Lemma 6.9 *The (matrix) kernel of the operator* $\rho_\mu(w) L_0 \rho_\nu(w)$ *is given by*

$$a(x+iy, z) = \sum_{s\in\mathbb{F}_q^*} \nu(sz) J(-s^{-1}(x+iy)^{-1}) j(-s^{-1}z^{-1}).$$

Proof By (5.10), for $f\in L(\mathbb{F}_q^*)$ we have that

$$\begin{aligned}[\rho_\mu(w)L_0\rho_\nu(w)f](x+iy) &= -\sum_{s+it\in\mathbb{F}_{q^2}^*}\mu(-(s+it))J(-(s+it)^{-1}(x+iy)^{-1})\\ &\quad\cdot[L_0\rho_\nu(w)f](s+it)\\ \text{(by (6.20) and } \theta=0) &= \sum_{z,s\in\mathbb{F}_q^*}\nu(sz)J(-s^{-1}(x+iy)^{-1})j(-s^{-1}z^{-1})f(z).\end{aligned}$$

□

Corollary 6.4

$$\sum_{u\in U_1}\rho_\mu(uw)L_0\rho_\nu(wu^{-1}) = q\sum_{\theta\in\mathbb{F}_q}\left[\sum_{t\in\mathbb{F}_q^*}\nu(-t^{-1})J(t(1-i\theta))j(t)\right]L_\theta.$$

Proof By Lemma 6.8 and with a as in Lemma 6.9, we have, for $f\in L(\mathbb{F}_q^*)$ and $x\neq 0$,

$$\begin{aligned}&\left[\sum_{\alpha\in\mathbb{F}_q}\rho_\mu\begin{pmatrix}1&\alpha\\0&1\end{pmatrix}\rho_\mu(w)L_0\,\rho_\nu(w)\rho_\nu\begin{pmatrix}1-\alpha\\0&1\end{pmatrix}f\right](x+iy)\\ &= qa\left(x+iy,\frac{x^2-\eta y^2}{x}\right)f\left(\frac{x^2-\eta y^2}{x}\right)\\ &= q\sum_{s\in\mathbb{F}_q^*}\nu(sx^{-1}(x^2-\eta y^2))J(-s^{-1}(x+iy)^{-1})\\ &\quad\cdot j(-s^{-1}x(x^2-\eta y^2)^{-1})f((x^2-\eta y^2)x^{-1})\\ \text{(by (6.20) with } \sigma=y/x) &= q\sum_{s\in\mathbb{F}_q^*}\nu(sx(1-\eta\sigma^2))J(-s^{-1}x^{-1}(1+i\sigma)^{-1})\\ &\quad\cdot j(-s^{-1}x^{-1}(1-\sigma^2\eta)^{-1})[L_\sigma f](x+iy)\\ &= q\sum_{\theta\in\mathbb{F}_q}\left[\sum_{t\in\mathbb{F}_q^*}\nu(-t^{-1})J(t(1-i\theta))j(t)\right][L_\theta f](x+iy),\end{aligned}$$

where, in the last equality, we have set $t=-s^{-1}x^{-1}(1-\eta\theta^2)^{-1}$ so that $-s^{-1}x^{-1}(1+i\theta)^{-1}=t(1-i\theta)$ and we have used, once again, (6.20). □

Theorem 6.6 *The operator* $\widetilde{L}$ *in* (6.44) *is equal to*

$$\frac{1}{q+1}L_0+\frac{q}{q+1}\sum_{\theta\in\mathbb{F}_q}\left[\sum_{t\in\mathbb{F}_q^*}\nu(-t^{-1})J(t(1-i\theta))j(t)\right]L_\theta.$$

Proof Using the Bruhat decomposition (cf. (3.32)) $G_1 = B_1 \bigsqcup (U_1 w B_1)$ and the fact that

$$\rho_\mu(b) L_0 = L_0 \rho_\nu(b) \qquad \text{for all } b \in B_1 \tag{6.46}$$

(see Corollary 6.1) we get

$$\begin{aligned}
\frac{1}{|G_1|} \sum_{g \in G_1} \rho_\mu(g) L_0 \rho_\nu(g^{-1}) &= \frac{1}{|G_1|} \Bigg[\sum_{b \in B_1} \rho_\mu(b) L_0 \rho_\nu(b^{-1}) + \\
&\quad + \sum_{u \in U_1} \sum_{b \in B_1} \rho_\mu(u) \rho_\mu(w) \rho_\mu(b) L_0 \rho_\nu(b^{-1}) \\
&\quad \cdot \rho_\nu(w) \rho_\nu(u^{-1}) \Bigg] \\
\text{(by (6.46))} &= \frac{|B_1|}{|G_1|} L_0 + \frac{|B_1|}{|G_1|} \sum_{u \in U_1} \rho_\mu(u) \rho_\mu(w) L_0 \rho_\nu(w) \rho_\nu(u^{-1})
\end{aligned}$$

so that the result follows immediately from Corollary 6.4. □

Remark 6.7 First of all, from (6.23) we deduce that

$$\widetilde{L}\delta_1 = \frac{1}{q+1}\delta_1 + \frac{q}{q+1} \sum_{\theta \in \mathbb{F}_q} \left[\sum_{t \in \mathbb{F}_q^*} \nu(-t^{-1}) J(t(1-i\theta)) j(t) \right] \delta_{(1-i\theta)^{-1}}$$

so that

$$[\widetilde{L}\delta_1]((1-i\sigma)^{-1}) = \frac{1}{q+1}\delta_{0,\sigma} + \frac{q}{q+1} \sum_{t \in \mathbb{F}_q^*} \nu(-t^{-1}) J(t(1-i\sigma)) j(t).$$

This coincides with (6.43) and yields a revisitation of the calculations in the preceding section. However, we are not able to compute the norm of $\widetilde{L}\delta_1$ or, more generally, of $\widetilde{L}f$, in order to normalize $\widetilde{L}$ and thus obtaining an *isometry*.

Problem 6.1 Compute the norm of $\widetilde{L}\delta_1$ or, more generally, of $\widetilde{L}f$, for $f \in L(\mathbb{F}_q^*)$.

A solution to the above problem would lead to a different approach to the computations in the preceding section. See also Remark 5.5 for a very similar problem.

Appendix A

A.1 On a Question of Ricci and Samanta

Recently, Ricci and Samanta [56, Corollary 3.3] proved the following result.

Theorem A.1 (Ricci and Samanta) *Let G be a locally compact Lie group, let K be a compact subgroup with G/K connected, and let τ be an irreducible K-representation such that the triple (G, K, τ) is multiplicity-free. Then (G, K) is a Gelfand pair.*

Note that the authors use the term "commutative" rather than "multiplicity-free". Then Ricci and Samanta pose the following natural question: is the same statement true outside of the realm of Lie groups? It turns out that, in the setting of finite groups, we are able to answer their question in the following striking manner:

Theorem A.2 $(\mathrm{GL}(2, \mathbb{F}_q), U, \chi)$ *is a multiplicity-free triple for every* non-trivial *additive character χ of the subgroup $U \cong \mathbb{F}_q$ of unipotent matrices* but $(\mathrm{GL}(2, \mathbb{F}_q), U)$ *is not a Gelfand pair, that is,* $(\mathrm{GL}(2, \mathbb{F}_q), U, \chi_0)$ *is not multiplicity-free if χ_0 is the trivial character.*

It was pointed out to us by one of the referees that the first result in the above theorem may be immediately deduced from the following result of Yokonuma [73] (see also [74]).

Theorem A.3 (Yokonuma) *Let G be a Chevalley group (i.e., the group of rational points of a split adjoint simple algebraic group) over $\mathbb{F}_q$. Let U be a maximal unipotent subgroup, and let χ be a character on U which is not trivial on any of the one-parameter subgroups X_a (here a is a simple root). Then the commutant of the induced representation $\mathrm{Ind}_U^G \chi$ is commutative and of dimension q^ℓ, where $\ell = rank(G)$. In other words, $\mathrm{Ind}_U^G \chi$ is multiplicity-free and splits as a sum of q^ℓ irreducible representations.*

T. Ceccherini-Silberstein et al., *Gelfand Triples and Their Hecke Algebras*, Lecture Notes in Mathematics 2267, https://doi.org/10.1007/978-3-030-51607-9

Proof of Theorem A.2 The first fact is proved in Theorem 3.9; the corresponding decomposition into irreducibles and the computation of the spherical functions are in [17, Sections 14.6 and 14.7]. The second fact is in [17, Exercise 14.5.10.(2)]. More precisely, we have (with $G = \mathrm{GL}(2, \mathbb{F}_q)$):

$$\mathrm{Ind}_U^G \chi_0 = \left(\bigoplus_{\psi \in \widehat{F_q^*}} \widehat{\chi}_\psi^0 \right) \bigoplus \left(\bigoplus_{\psi \in \widehat{F_q^*}} \widehat{\chi}_\psi^1 \right) \bigoplus 2 \left(\bigoplus_{\{\psi_1, \psi_2\}} \widehat{\chi}_{\psi_1, \psi_2} \right), \tag{A.1}$$

where the last sum runs over all two-subsets of $\widehat{F_q^*}$. In order to prove the decomposition (A.1), we do not follow the proof indicated in the exercise in our monograph, because it is based on a detailed analysis of the induced representations involving also those of the affine group, but we follow the lines of the proof of Theorem 5.1. To this end, we compute the restriction to U of the irreducible representations of $\mathrm{GL}(2, \mathbb{F}_q)$. First of all, note that if $b \neq 0$ then

$$\begin{pmatrix} b^{-1} & 0 \\ 0 & 1 \end{pmatrix} \begin{pmatrix} 1 & b \\ 0 & 1 \end{pmatrix} \begin{pmatrix} b & 0 \\ 0 & 1 \end{pmatrix} = \begin{pmatrix} 1 & 1 \\ 0 & 1 \end{pmatrix},$$

that is, the unipotent elements

$$\begin{pmatrix} 1 & b \\ 0 & 1 \end{pmatrix} \quad \text{and} \quad \begin{pmatrix} 1 & 1 \\ 0 & 1 \end{pmatrix},$$

are conjugate, Therefore, from the character table of $\mathrm{GL}(2, \mathbb{F}_q)$ (cf. [17, Table 14.2, Section 4.9]) we get, for all $b \in \mathbb{F}_q$:

- $\widehat{\chi}_\psi^0 \begin{pmatrix} 1 & b \\ 0 & 1 \end{pmatrix} \equiv \widehat{\chi}_\psi^0 \begin{pmatrix} 1 & 1 \\ 0 & 1 \end{pmatrix} = \psi(1) = 1;$
- $\chi^{\widehat{\chi}_\psi^1} \begin{pmatrix} 1 & b \\ 0 & 1 \end{pmatrix} = \begin{cases} 0 & \text{if } b \neq 0 \\ q & \text{if } b = 0; \end{cases}$
- $\chi^{\widehat{\chi}_{\psi_1, \psi_2}} \begin{pmatrix} 1 & b \\ 0 & 1 \end{pmatrix} = \begin{cases} \psi_1(1)\psi_2(1) = 1 & \text{if } b \neq 0 \\ (q+1)\psi_1(1)\psi_2(1) = (q+1) & \text{if } b = 0; \end{cases}$
- $\chi^{\rho_\nu} \begin{pmatrix} 1 & b \\ 0 & 1 \end{pmatrix} = \begin{cases} -\nu(1) = -1 & \text{if } b \neq 0 \\ (q-1)\nu(1) = (q-1) & \text{if } b = 0. \end{cases}$

By Frobenius reciprocity, the multiplicity of $\widehat{\chi}_\psi^0$ in $\mathrm{Ind}_U^G \chi_0$ is equal to the multiplicity of χ_0 in $\mathrm{Res}_U^G \widehat{\chi}_\psi^0$, that is to

$$\frac{1}{q} \sum_{b \in \mathbb{F}_q} 1 = 1.$$

This proves the first block in (A.1). Similarly, for $\widehat{\chi}^1_\psi$ we get:

$$\frac{1}{q}\left(\sum_{b\in\mathbb{F}_q^*} 0+q\right) = 1,$$

and this proves the second block. For $\widehat{\chi}_{\psi_1,\psi_2}$ we get:

$$\frac{1}{q}\left(\sum_{b\in\mathbb{F}_q^*} 1+(q+1)\right) = \frac{q-1+q+1}{q} = 2,$$

so that the third block is proved, showing, in particular, that multiplicities occur. Finally, for ρ_ν we get:

$$\frac{1}{q}\left(\sum_{b\in\mathbb{F}_q^*} (-1)+(q-1)\right) = \frac{-(q-1)+(q-1)}{q} = 0,$$

showing that no cuspidal representations appear in the decomposition (A.1).

A.2 The Gelfand Pair $(\mathrm{GL}(2, \mathbb{F}_{q^2}), \mathrm{GL}(2, \mathbb{F}_q))$

Akihiro Munemasa [51], after reading a preliminary version of the present monograph, pointed out to us that $(\mathrm{GL}(2, \mathbb{F}_{q^2}), \mathrm{GL}(2, \mathbb{F}_q))$ is a Gelfand pair. This is due to Gow [37, Theorem 3.6] who proved a more general result, namely, that $(\mathrm{GL}(n, \mathbb{F}_{q^2}), \mathrm{GL}(n, \mathbb{F}_q))$ is a Gelfand pair, for any $n \geq 1$. Gow's result was generalized by Henderson [41] who, using Lusztig's crucial work on character sheaves, showed that $(G(\mathbb{F}_{q^2}), G(\mathbb{F}_q))$ is a Gelfand pair for any connected reductive algebraic group G, and found an effective algorithm for computing the corresponding spherical functions. This shares a strong similarity with the results of Bannai, Kawanaka, and Song [2], where the Gelfand pair $(\mathrm{GL}_{2n}(\mathbb{F}_q), \mathrm{Sp}_{2n}(\mathbb{F}_q))$ is analyzed (here Sp stands for the *symplectic group*).

A proof of Gow's theorem (for $n = 2$) can be directly deduced from our computations in our monograph [17, Section 14.11], where we studied induction from $\mathrm{GL}(2, \mathbb{F}_q)$ to $\mathrm{GL}(2, \mathbb{F}_{q^m})$ for $m \geq 2$. Since the relative decompositions are left as a terrific set of exercises and the case $m = 2$ it is not well specified, we now complement Proposition 6.1 by presenting the formulas for the induction of the other representations (the one-dimensional and the parabolic representations).

We set $G = \mathrm{GL}(2, \mathbb{F}_q)$ and $G_m = \mathrm{GL}(2, \mathbb{F}_{q^m})$. Moreover, from Sect. 5.3 we use the notation (5.4) and for $\xi \in \widehat{\mathbb{F}^*_{q^2}}$ we set $\xi^\sharp = \mathrm{Res}^{\mathbb{F}^*_{q^2}}_{\mathbb{F}^*_q} \xi$. Then the first formula on page 537 of our monograph (namely, the decomposition of the induced representation $\mathrm{Ind}_G^{G_m} \widehat{\chi}^0_\psi$) for $m = 2$ becomes:

$$\mathrm{Ind}_G^{G_2} \widehat{\chi}^0_\psi = \left(\bigoplus_{\xi^\sharp = \psi} \widehat{\chi}^0_\xi \right) \oplus \left(\bigoplus_{\xi^\sharp = \psi} \widehat{\chi}^1_\xi \right) \oplus \left(\bigoplus_{\xi_1^\sharp = \xi_2^\sharp = \psi} \widehat{\chi}_{\xi_1, \xi_2} \right) \oplus \left(\bigoplus_{\overline{\xi}_1 \xi_2 = \Psi} \widehat{\chi}_{\xi_1, \xi_2} \right).$$

This is *multiplicitiy–free*. To show this, it suffices to prove that the third and the fourth block have no common summands. Otherwise, if $\widehat{\chi}_{\xi_1, \xi_2}$ were in both blocks, we would have $\overline{\xi}_1 \xi_2 = \Psi$ and $\xi_1^\sharp = \psi$. Then

$$\xi_1(z)\overline{\xi}_1(z) = \xi_1(z)\xi_1(\overline{z}) = \xi_1(z\overline{z}) = \psi(z\overline{z}) = \Psi(z) = \overline{\xi}_1(z)\xi_2(z) \tag{A.2}$$

for all $z \in \mathbb{F}^*_{q^2}$, and therefore $\xi_1 = \xi_2$, a contradiction. In particular, if ψ is the trivial (multiplicative) character of $\mathbb{F}_q$, so that $\widehat{\chi}^0_\psi$ equals ι_G, the trivial representation of G, we have that $\mathrm{Ind}_G^{G_2} \chi^0_\psi$ decomposes without multiplicities. We thus obtain

Theorem A.4 (Gow) $(\mathrm{GL}(2, \mathbb{F}_{q^2}), \mathrm{GL}(2, \mathbb{F}_q))$ *is a Gelfand pair.*

Returning back to our computations, the second formula on page 537 of our monograph (namely, the decomposition of the induced representation $\mathrm{Ind}_G^{G_m} \widehat{\chi}^1_\psi$) for $m = 2$ becomes:

$$\mathrm{Ind}_G^{G_2} \widehat{\chi}^1_\psi = \left(\bigoplus_{(\xi^\sharp)^2 = \psi^2} \widehat{\chi}^1_\xi \right) \oplus \left(\bigoplus_{\substack{(\xi_1\xi_2)^\sharp = \psi^2 \\ \overline{\xi}_1 \xi_2 \neq \Psi}} \widehat{\chi}_{\xi_1, \xi_2} \right)$$
$$\oplus \left(\bigoplus_{\xi_1^\sharp = \xi_2^\sharp = \psi} \widehat{\chi}_{\xi_1 \xi_2} \right) \oplus \left(\bigoplus_{\nu^\sharp = \psi^2} \rho_\nu \right).$$

Now multiplicities do appear! Indeed, each representation in the third block is also in the second block. For, if $\xi_1^\sharp = \xi_2^\sharp = \psi$ then also $(\xi_1\xi_2)^\sharp = \psi^2$ (and $\overline{\xi}_1\xi_2 \neq \Psi$, as proved above, cf. (A.2)).

Finally, Formula (14.64) in [17] for $m = 2$ becomes:

$$\mathrm{Ind}_G^{G_2} \widehat{\chi}_{\psi_1,\psi_2} = \left(\bigoplus_{(\xi^\sharp)^2=\psi_1\psi_2} \widehat{\chi}^1_\xi \right) \bigoplus \left(\bigoplus_{(\xi_1\xi_2)^\sharp=\psi_1\psi_2} \widehat{\chi}_{\xi_1,\xi_2} \right)$$
$$\bigoplus \left(\bigoplus_{\substack{\xi_1^\sharp=\psi_1 \\ \xi_2^\sharp=\psi_2}} \widehat{\chi}_{\xi_1\xi_2} \right) \bigoplus \left(\bigoplus_{\nu^\sharp=\psi_1\psi_2} \rho_\nu \right).$$

Once more, multiplicities do occur, since the third block is clearly contained in the second. In conclusion, we have the following strengthening of Theorem A.4:

Theorem A.5 *The decomposition of the induced representation* $\mathrm{Ind}_G^{G_2}\theta$ *is multiplicity-free if and only if* θ *is cuspidal or one-dimensional. If* θ *is parabolic, then some subrepresentations in* $\mathrm{Ind}_G^{G_2}\theta$ *appear with multiplicity* 2.

A.3 On some Questions of Dunkl

Charles Dunkl [26], after reading a preliminary version of the manuscript, pointed out to us interesting connections with other similar constructions and asked a few questions in relation with these.

1. **Algebras of conjugacy-invariant functions.** The first one is related to the work of I.I. Hirschman [42] which involves two subgroups $K \leq H \leq G$ (with H contained in the normalizer of K) of a finite group G, and yields a subalgebra $L(G, H, K)$ of $L(G)$. We thus recall from [13, Chapter 2] a generalization of the theory of subgroup-conjugacy-invariant algebras due to A. Greenhalgh [39] and that, following Diaconis [25], we called Greenhalgebras. These were also considered by Brender [6] (but Greenhalgh considered a more general case).

Let G be a finite group and suppose that K and H are two subgroups of G, with $K \trianglelefteq H \leq G$ (thus, in particular, H is contained in the normalizer of K, as in [42]). Then, the *Greenhalgebra* associated with G, H, and K is the subalgebra $\mathscr{G}(G, H, K)$ of $L(G)$ consisting of all functions that are both H-conjugacy-invariant and bi-K-invariant (cf. [13, Section 2.1.3]):

$$\mathscr{G}(G, H, K) = \{f \in L(G) : f(h^{-1}gh) = f(g) \text{ and } f(k_1 g k_2)$$
$$= f(g), \forall g \in G,\ h \in H,\ k_1, k_2 \in K\}.$$

Set $\widetilde{H} = \{(h, h) : h \in H\} \leq G \times G$ and

$$B = (K \times \{1_G\})\widetilde{H} = \{(kh, h) : k \in K \text{ and } h \in H\} \leq H \times H.$$

Given an irreducible representation $\theta \in \widehat{B}$ we say that the quadruple $(G, H, K; \theta)$ is *multiplicity-free* provided the induced representation $\mathrm{Ind}_B^{G\times H}\theta$ decomposes without multiplicities. Let us also set $\widehat{H}_K = \{\rho \in \widehat{H} : \mathrm{Res}^H_K \rho = (\dim \rho)\iota_K\}$ (where ι_K is the trivial representation of K). We note (cf. [13, Lemma 2.1.15]) that $\mathscr{G}(G, H, K)$ is isomorphic to the algebra

$$\begin{aligned} {}^B L(G \times H)^B = \{f \in L(G \times H) : \ & f(b_1(g, h)b_2) \\ &= f(g, h) \text{ for all } b_1, b_2 \in B \text{ and } g \in G, h \in H\} \end{aligned}$$

of all bi-B-invariant functions on $G \times H$.

When $\theta = \iota_B$ is the trivial representation of B we have (cf. [13, Theorem 2.1.19]):

Theorem A.6 *With the above notation, the following conditions are equivalent:*

(a) *the quadruple $(G, H, K; \iota_B)$ is multiplicity-free;*
(b) *the Greenhalgebra $\mathscr{G}(G, H, K)$ is commutative;*
(c) *$(G \times H, B)$ is a Gelfand pair;*
(d) *for every $\sigma \in \widehat{G}$ and $\rho \in \widehat{H}_K$, the multiplicity of σ in $\mathrm{Ind}_H^G \rho$ is ≤ 1.*

Note that when $K = \{1_G\}$, then $B = \widetilde{H}$ and $\mathscr{G}(G, H, K)$ is simply the subalgebra $\mathscr{C}(G, H) = \{f \in L(G) : f(h^{-1}gh) = f(g) \text{ for all } g \in G \text{ and } h \in H\}$ of all H-conjugacy-invariant functions on G (cf. [13, Section 2.1.1]). Moreover, Theorem A.6 yields (cf. [13, Theorem 2.1.10]):

Theorem A.7 *The following conditions are equivalent:*

(a) *the quadruple $(G, H, \{1_G\}; \iota_{\widetilde{H}})$ is multiplicity-free;*
(b) *the algebra $\mathscr{C}(G, H)$ is commutative;*
(c) *$(G \times H, \widetilde{H})$ is a Gelfand pair;*
(d) *H is a multiplicity-free subgroup of G.*

Returning back to a general irreducible representation $\theta \in \widehat{B}$, motivated by Dunkl's question, we pose the following:

Problem A.1 Given a quadruple $(G, H, K; \theta)$ as above, define a Hecke-type Greenhalgebra $\mathscr{H}\mathscr{G}(G, H, K; \theta)$ in a such a way that (i) $\mathscr{H}\mathscr{G}(G, H, K; \iota_B) = \mathscr{G}(G, H, K)$ (that is, when $\theta = \iota_B$ is the trivial representation, then the Hecke-type Greenhalgebra coincides with the Greenhalgebra of the triple (G, H, K)) and (ii) $\mathscr{H}\mathscr{G}(G, H, K; \theta)$ is commutative if and only if the quadruple $(G, H, K; \theta)$ is multiplicity-free.

We shall try to address this problem in a future paper.

2. **Positive-definite functions.** Let G be a finite group. A function $\phi\colon G \to \mathbb{C}$ is said to be *positive-definite* (or *of positive type*) if equivalently (see [33, Section 3.3], [55, Capitolo VII, Sezione 7]):
   (a) $\sum_{g,h\in G} \phi(h^{-1}g) f(g)\overline{f(h)} \geq 0$ for all $f \in L(G)$;
   (b) $\sum_{i,j=1}^{n} c_i\overline{c_j}\phi(g_j^{-1}g_i) \geq 0$ for all $c_1, c_2, \ldots, c_n \in \mathbb{C}$, $g_1, g_2, \ldots, g_n \in G$, and $n \geq 1$;
   (c) there exists a (unitary) representation (σ_ϕ, V_ϕ) of G and a (cyclic) vector $v_\phi \in V_\phi$ such that $\phi(g) = \langle \sigma_\phi(g)v_\phi, v_\phi\rangle_{V_\phi}$ for all $g \in G$.

Suppose that (G, K) is a Gelfand pair. Let $\phi \in L(G)$ be a function of positive type. Then ϕ is bi-K-invariant if and only if $\sigma_\phi(k)v_\phi = v_\phi$ for all $k \in K$; additionally, ϕ is spherical if and only if σ_ϕ is irreducible. Moreover, the spherical functions of positive type together with the zero function are exactly the extremal points of the (convex) set of all functions of positive type.

In our setting, the spherical functions ϕ^σ, $\sigma \in J$ as in (4.2.4) are positive-definite.

Recall that the Hecke algebra $\mathscr{H}(G, K, \psi) \leq L(G)$ is isomorphic, via the isomorphism S_v^{-1} (cf. Theorem 3.9) to the Hecke algebra $\widetilde{\mathscr{H}}(G, K, \theta)$ which consists of $\mathrm{End}_G(V_\theta)$-valued functions on G. Then the matrix-valued functions $S_v^{-1}(\phi^\sigma) \in \widetilde{\mathscr{H}}(G, K, \theta)$ are positive-definite in the sense of Dunkl (cf. [29, Definition 5.1]: with $H = K$).

A comment about positive-definite functions on finite groups: we thank Charles Dunkl for this interesting historical information. In the mid-Seventies, L.L. Scott, while involved in the search for new simple finite groups, computed possible value tables for the spherical functions of homogeneous spaces of rank 3. He showed his computations to his colleague Dunkl at the University of Virginia who checked them and found that, as this positivity condition for the functions involved therein was missing (cf. [29, Section 5]), the corresponding tables could not be valid (so that this possibility had to be eliminated). Scott wrote this up and dubbed it the *Krein condition* (this positivity condition in finite permutation groups is perfectly analogous to a condition obtained in harmonic analysis by the Soviet mathematician M.G. Krein) [62, 63].

3. **Finite hypergroups.** A finite *(algebraic) hypergroup* (see, e.g., [19]) is a pair $(X, *)$, where X is a nonempty finite set equipped with a multi-valued map, called *hyperoperation* and denoted $*$, from $X \times X$ to $\mathscr{P}^*(X)$, the set of all nonempty subsets of X, satisfying the following properties:
   (i) $(x * y) * z = x * (y * z)$ for all $x, y, z \in X$ (*associative property*);
   (ii) $x * X = X * x = X$ for all $x \in X$ (*reproduction property*),
   where for subsets $Y, Z \subset X$ one defines $Y * Z = \{y * z : y \in Y, z \in Z\} \subset X$.

If, in addition one has

(iii) $x * y = y * x$ for all $x, y \in X$ (*commutative property*)
one says that $(X, *)$ is commutative. Also, an element $e \in X$ is called a *unit* provided

(iv) $x \in (e * x) \cap (x * e)$ for all $x \in X$.

Given a finite set X, we denote by $L^1(X)$ (resp. $L^1_+(X)$) the space of all function $f \in L(X)$ such that $\sum_{x \in X} f(x) = 1$ (resp. $f \in L^1(X)$ such that $f \geq 0$).

A finite *functional hypergroup* (cf. [27, 28, 45]) is a pair (X, λ) where X is a nonempty finite set and $\lambda \colon X \times X \to L^1_+(X)$ satisfies the following properties:

(i') $(\mu *_\lambda \nu) *_\lambda \xi = \mu *_\lambda (\nu *_\lambda \xi)$ for all $\mu, \nu, \xi \in L^1_+(X)$ (*associative property*),

(ii') $\cup_{y \in Y} \text{supp}(\lambda(x, y)) = \cup_{y \in Y} \text{supp}(\lambda(y, x)) = X$ for all $x \in X$ (*reproduction property*),
where, $\delta_x \in L^1_+(X)$ is the Dirac delta at $x \in X$, $\text{supp}(f) = \{x \in X : f(x) \neq 0\}$ is the *support* of $f \in L^1(X)$, and $*_\lambda \colon L^1_+(X) \times L^1_+(X) \to L^1_+(X)$ is the (bilinear) product defined by

$$\mu *_\lambda \nu = \sum_{x, y \in X} \mu(x) \mu(y) \lambda(x, y) \tag{A.3}$$

for all $\mu = \sum_{x \in X} \mu(x) \delta_x$ and $\nu = \sum_{y \in X} \nu(y) \delta_y$ in $L^1_+(X)$. Note that (A.3) is equivalent to

$$\delta_x *_\lambda \delta_y = \lambda(x, y) \tag{A.4}$$

for all $x, y \in X$.

If, in addition, one has

(iii') $\lambda(x, y) = \lambda(y, x)$ for all $x, y \in X$ (*commutative property*)
one says that (X, λ) is commutative. Finally, an element $e \in X$ such that

(iv') $\lambda(x, e) = \lambda(e, x) = \delta_x$ for all $x \in X$, is called a unit of the functional hypergroup X.

A (finite) *weak functional hypergroup* is defined verbatim except that one replaces $L^1_+(X)$ by $L^1(X)$, that is, one drops the condition $\lambda(x, y) \geq 0$ for all $x, y \in X$. Clearly, every functional hypergroup is a weak functional hypergroup.

Example A.1

1. Every finite algebraic hypergroup is a finite functional hypergroup. Indeed, given a finite algebraic hypergroup $(X, *)$ one may set $\lambda(x, y) = \frac{1}{|x*y|} \sum_{z \in x*y} \delta_z$ for all $x, y \in X$. Clearly, an algebraic hypergroup is commutative (resp. has a unit) if and only if the associated functional hypergroup is (resp. has). Vice versa, with any functional hypergroup (X, λ) one may associate an algebraic hypergroup by setting $x * y = \text{supp}(\lambda(x, y))$ for all $x, y \in X$ (cf. [27, Proposition 1.4]).

2. Every finite group (resp. Abelian group) $(G, \cdot)$ is an algebraic hypergroup (resp. an Abelian hypergroup). This follows immediately after defining $x * y = \{x \cdot y\}$. Indeed, (i) follows from associativity of the group operation, while (ii) is a consequence of the fact that left and right multiplication by a fixed group element constitutes a permutation of the group. Moreover the identity element of G serves as a unit for the hypergoup.
3. Let G be a finite group. Then $X = \widehat{G}$ is an algebraic hypergroup after setting $x * y = \{z \in X : z \preceq x \otimes y\}$ for all $x, y \in X$. Equivalently, X is a functional hypergroup by setting $\lambda(x, y)(z) = \frac{\dim V_z}{\dim V_x \dim V_y} \dim \mathrm{Hom}_G(z, x \otimes y)$ (in other words, $\lambda(x, y)(z) = \frac{\dim V_z}{\dim V_x \dim V_y} \mu_{xyz}$, where $x \otimes y = \sum_{z \in X} \mu_{xyz} z$) for all $x, y, z \in X$. Moreover, the trivial representation $\iota_G \in X$ serves as a unit for the hypergoup.
4. Let G be a finite group and let X denote the set of all conjugacy classes of G. Given $x \in X$ we denote by $f_x = \frac{1}{|x|} \sum_{g \in x} \delta_g \in L^1_+(X)$ the normalized characteristic function of $x \subseteq G$. It is well known that $\{f_x : x \in X\}$ constitutes a base for the subspace $\mathscr{C}(G) = \{f \in L(G) : f(h^{-1}gh) = f(g)$ for all $g, h \in G\}$ of *conjugacy-invariant* functions on G and that, indeed, $\mathscr{C}(G)$ is a subalgebra of $L(G)$. We may thus define a functional hypergroup (X, λ) by setting $\lambda(x, y) = f_x * f_y \in L^1_+(G)$ (where $*$ is the usual convolution) for all $x, y \in X$. Moreover, the conjugacy class of the identity element of G serves as a unit for the hypergoup.
5. Let G be a finite group and let $K \leq G$ be a subgroup. Denote by $X = K\backslash G/K = \{KgK : g \in G\}$ the set of all double K-cosets of G. For $x \in X$ denote by $f_x = \frac{1}{|x|} \sum_{g \in x} \delta_g \in L^1_+(X)$ the normalized characteristic function of $x \in X$. It is well known that $\{f_x : x \in X\}$ constitutes a basis for the subspace ${}^K L(G)^K = \{f \in L(G) : f(k_1^{-1} g k_2) = f(g)$ for all $g \in G$ and $k_1, k_2 \in K\}$ of *bi-K-invariant* functions on G and that, indeed, ${}^K L(G)^K$ is a subalgebra of $L(G)$. We may thus define a functional hypergroup (X, λ) by setting $\lambda(x, y) = f_x * f_y \in L^1_+(G)$ (where $*$ is the usual convolution) for all $x, y \in X$. The double coset $K = K\{1_G\}K$ serves as a unit for the hypergoup. Finally, (X, λ) is commutative if and only if ${}^K L(G)^K$ is commutative, equivalenlty, if and only if (G, K) is a Gelfand pair.
6. In the setting of the present paper (see also [17, Section 13.2]), let G be a finite group, $K \leq G$ a subgroup, and suppose that $\chi \in \widehat{K}$ is one-dimensional. Consider the Hecke algebra

$$\mathscr{H}(G, K, \chi) = \left\{ f \in L(G) : f(k_1 g k_2) = \overline{\chi(k_1 k_2)} f(g), \text{ for all } g \in G, k_1, k_2 \in K \right\}$$

(when $\chi = \iota_K$, the trivial representation of K, this is nothing but ${}^K L(G)^K$, the algebra of bi-K-invariant, discussed in (4) above). Let $\mathscr{S}$ be a set of representatives for the set $K\backslash G/K$ of double K-cosets and set

$$X = \mathscr{S}_0 = \{s \in \mathscr{S} : \chi(x) = \chi(s^{-1}xs) \text{ for all } x \in K \cap sKs^{-1}\}.$$

Then the functions $a_x \in L(G)$ defined by

$$a_x(g) = \begin{cases} \frac{1}{|K|}\overline{\chi(k_1)\chi(k_2)} & \text{if } g = k_1 x k_2 \text{ for some } k_1, k_2 \in K \\ 0 & \text{if } g \notin KxK \end{cases}$$

for all $g \in G$ and $x \in X$, constitute a basis (called the *Curtis and Fossum* basis, [20]) for $\mathscr{H}(G, K, \chi)$ and the numbers μ_{xyz} such that

$$a_x * a_y = \sum_{z \in X} \mu_{xyz} a_z \tag{A.5}$$

for all $x, y, z \in X$, are called the *structure constants* of $\mathscr{H}(G, K, \chi)$. Denoting by $f_x = \frac{1}{|KxK|} a_x \in L^1(G)$ the corresponding normalization, we have, for all $x, y \in X$,

$$f_x * f_y = \sum_{z \in X} \mu'_{xyz} f_z,$$

where $\mu'_{xyz} = \frac{|KzK|}{|KxK| \cdot |KyK|} \mu_{xyz} \in \mathbb{C}$. We may thus define a weak functional hypergroup (X, λ) by setting $\lambda(x, y) = \sum_{z \in X} \mu'_{xyz} f_z$ for all $x, y \in X$.

References

1. E. Bannai, T. Ito, *Algebraic Combinatorics* (Benjamin, Menlo Park, 1984)
2. E. Bannai, N. Kawanaka, S.-Y. Song, The character table of the Hecke algebra $\mathscr{H}(\mathrm{GL}_{2n}(\mathbb{F}_q); \mathrm{Sp}_{2n}(\mathbb{F}_q))$. J. Algebra **129**, 320–366 (1990)
3. E. Bannai, H. Tanaka, The decomposition of the permutation character $1_{\mathrm{GL}(n,q^2)}^{\mathrm{GL}(2n,q)}$. J. Algebra **265**(2), 496–512 (2003)
4. E. Bannai, H. Tanaka, Appendix: on some Gelfand pairs and commutative association schemes. Jpn. J. Math. **10**(1), 97–104 (2015)
5. Ya.G. Berkovich, E.M. Zhmud', *Characters of Finite Groups*. Part 1. Translations of Mathematical Monographs, vol. 172 (American Mathematical Society, Providence, 1998)
6. M. Brender, A class of Schur algebras. Trans. Am. Math. Soc. **248**(2), 435–444 (1979)
7. D. Bump, *Lie Groups*. Graduate Texts in Mathematics, vol. 225 (Springer, New York, 2004)
8. D. Bump, D. Ginzburg, Generalized Frobenius–Schur numbers. J. Algebra **278**(1), 294–313 (2004)
9. T. Ceccherini-Silberstein, A. Machì, F. Scarabotti, F. Tolli, Induced representations and Mackey theory. J. Math. Sci. (N.Y.) **156**(1), 11–28 (2009)
10. T. Ceccherini-Silberstein, F. Scarabotti, F. Tolli, Finite Gelfand pairs and their applications to probability and statistics. J. Math. Sci. (N.Y.) **141**(2), 1182–1229 (2007)
11. T. Ceccherini-Silberstein, F. Scarabotti, F. Tolli, *Harmonic Analysis on Finite Groups: Representation Theory, Gelfand Pairs and Markov Chains*. Cambridge Studies in Advanced Mathematics, vol. 108 (Cambridge University Press, Cambridge, 2008)
12. T. Ceccherini-Silberstein, F. Scarabotti, F. Tolli, Clifford theory and applications. J. Math. Sci. (N.Y.) **156**(1), 29–43 (2009)

13. T. Ceccherini-Silberstein, F. Scarabotti, F. Tolli, *Representation Theory of the Symmetric Groups: The Okounkov–Vershik Approach, Character Formulas, and Partition Algebras*. Cambridge Studies in Advanced Mathematics, vol. 121 (Cambridge University Press, Cambridge, 2010)
14. T. Ceccherini-Silberstein, F.Scarabotti, F.Tolli, *Representation Theory and Harmonic Analysis of Wreath Products of Finite Groups*. London Mathematical Society Lecture Note Series, vol. 410 (Cambridge University Press, Cambridge, 2014)
15. T. Ceccherini-Silberstein, F. Scarabotti, F. Tolli, Mackey's theory of τ-conjugate representations for finite groups. Jpn. J. Math. **10**(1), 43–96 (2015)
16. T. Ceccherini-Silberstein, F. Scarabotti, F. Tolli, Mackey's criterion for subgroup restriction of Kronecker products and harmonic analysis on Clifford groups. Tohoku Math. J. **67**(4), 553–571 (2015)
17. T. Ceccherini-Silberstein, F.Scarabotti, F.Tolli, *Discrete Harmonic Analysis: Representations, Number Theory, Expanders, and the Fourier Transform*. Cambridge Studies in Advanced Mathematics, vol. 172 (Cambridge University Press, Cambridge, 2018)
18. T. Ceccherini-Silberstein, F. Scarabotti, F. Tolli, Induced representations, Clifford theory, Mackey obstruction, and applications. Work in progress
19. P. Corsini, V. Leoreanu, *Applications of Hyperstructure Theory* (Springer, Berlin, 2003)
20. C.W. Curtis, T.V Fossum, On centralizer rings and characters of representations of finite groups. Math. Z. **107**, 402–406 (1968)
21. Ch.W. Curtis, I. Reiner, *Representation Theory of Finite Groups and Associative Algebras*. Reprint of the 1962 original. Wiley Classics Library. A Wiley-Interscience Publication (Wiley, New York, 1988)
22. Ch.W. Curtis, I. Reiner, *Methods of Representation Theory. with Applications to Finite Groups and Orders*. Pure and Applied Mathematics, vol. I (Wiley, New York 1981)
23. R. de la Madrid, The role of the rigged Hilbert space in quantum mechanics. Eur. J. Phys. **26**(2), 287–312 (2005)
24. P. Diaconis, *Groups Representations in Probability and Statistics* (IMS, Hayward, 1988)
25. P. Diaconis, Patterned matrices. Matrix theory and applications (Phoenix, AZ, 1989), in *Proceedings of the Symposium in Applied Mathematics*, vol. 40 (American Mathematical Society, Providence, 1990), pp. 37–58
26. Ch.F. Dunkl, *Private Communication*
27. Ch.F. Dunkl, The measure algebra of a locally compact hypergroup. Trans. Am. Math. Soc. **179**, 331–348 (1973)
28. Ch.F. Dunkl, Structure hypergroups for measure algebras. Pac. J. Math. **47**, 413–425 (1973)
29. Ch.F. Dunkl, Spherical functions on compact groups and applications to special functions, in *(Convegno sull'Analisi Armonica e Spazi di Funzioni su Gruppi Localmente Compatti, INDAM, Rome, 1976)*. Symposia Mathematica, vol. XXII (Academic Press, London, 1977), pp. 145–161
30. Ch.F. Dunkl, Orthogonal functions on some permutation groups, *Proceedings of the Symposium in Pure Mathematics*, vol. 34 (American Mathematical Society, Providence, 1979), pp. 129–147
31. J. Faraut, Analyse harmonique sur les paires de Guelfand et les espaces hyperboliques, CIMPA Lecture Notes (1980)
32. J.M.G. Fell and R.S. Doran, Representations of *-algebras, locally compact groups, and Banach *-algebraic bundles, vol. 2, in *Banach *-Algebraic Bundles, Induced Representations, and the Generalized Mackey Analysis*. Pure and Applied Mathematics, vol. 126 (Academic Press, Boston, 1988)
33. G.B. Folland *A Course in Abstract Harmonic Analysis*. Studies in Advanced Mathematics (CRC Press, Boca Raton, 1995)
34. I.M. Gelfand, M.I. Graev, Construction of irreducible representations of simple algebraic groups over a finite field. Dokl. Akad. Nauk SSSR **147**, 529–532 (1962)
35. I.M. Gelfand, N.J. Vilenkin, *Generalized Functions, vol. 4: Some Applications of Harmonic Analysis. Rigged Hilbert Spaces* (Academic Press, New York, 1964)

36. R. Godement, A theory of spherical functions. I. Trans. Am. Math. Soc. **73**, 496–556 (1952)
37. R. Gow, Two multiplicity-free permutations of the general linear group GL(n; q^2). Math. Z. **188**, 45–54 (1984)
38. J.A. Green, The characters of the finite general linear groups. Trans. Am. Math. Soc. **80**, 402–447 (1955)
39. A.S. Greenhalgh, Measure on groups with subgroups invariance properties. Technical report No. 321, Department of Statistics, Stanford University, 1989
40. P. de la Harpe, *Private communication*
41. A. Henderson, Spherical functions of the symmetric space $G(\mathbb{F}_{q^2})/G(\mathbb{F}_q)$. Represent. Theory **5**, 581–614 (2001)
42. I.I. Hirschman Jr., Integral equations on certain compact homogeneous spaces. SIAM J. Math. Anal. **3**, 314–343 (1972)
43. B. Huppert, *Character Theory of Finite Groups*. De Gruyter Expositions in Mathematics, vol. 25 (Walter de Gruyter, New York, 1998)
44. I.M. Isaacs, *Character Theory of Finite Groups*. Corrected reprint of the 1976 original [Academic Press, New York] (Dover Publications, New York, 1994)
45. R.J. Jewett, Spaces with an abstract convolution of measures. Adv. Math. **18**(1), 1–101 (1975)
46. I.G. Macdonald, *Symmetric Functions and Hall Polynomials*, 2nd edn. With contributions by A. Zelevinsky. Oxford Mathematical Monographs (Oxford Science Publications, The Clarendon Press, Oxford University Press, New York, 1995)
47. R.D. Martín, F. Levstein, Spherical analysis on homogeneous vector bundles of the 3-dimensional Euclidean motion group. Monatsh. Math. **185**(4), 621–649 (2018)
48. A. Mihailovs, The orbit method for finite groups of nilpontency class two of odd order. arXiv.org: math.RT/0001092
49. H. Mizukawa, Twisted Gelfand pairs of complex reflection groups and r-congruence properties of Schur functions. Ann. Comb. **15**(1), 119–125 (2011)
50. H. Mizukawa, Wreath product generalizations of the triple (S_{2n}, H_n, ϕ) and their spherical functions. J. Algebra **334**, 31–53 (2011)
51. A. Munemasa, *Private communication*
52. M.A. Naimark, A.I. Stern, *Theory of Group Representations* (Springer, New York, 1982)
53. I. Piatetski-Shapiro, *Complex Representations of GL(2,K) for Finite Fields K*. Contemporary Mathematics, vol. 16 (American Mathematical Society, Providence, 1983)
54. A. Okounkov, A.M. Vershik, A new approach to representation theory of symmetric groups. Selecta Math. (N.S.) **2**(4), 581–605 (1996)
55. F. Ricci, *Analisi di Fourier non Commutativa*, Class Notes SNS, 2018
56. F. Ricci, A. Samanta, Spherical analysis on homogeneous vector bundles. Adv. Math. **338**, 953–990 (2018)
57. J. Saxl, On multiplicity-free permutation representations, in *Finite Geometries and Designs*. London Mathematical Society Lecture Note Series, vol. 48 (Cambridge University Press, Cambridge, 1981), pp. 337–353
58. F. Scarabotti, F. Tolli, Harmonic analysis on a finite homogeneous space. Proc. Lond. Math. Soc. **100**(2), 348–376 (2010)
59. F. Scarabotti, F. Tolli, Fourier analysis of subgroup-conjugacy invariant functions on finite groups. Monatsh. Math. **170**, 465–479 (2013)
60. F. Scarabotti, F. Tolli, Hecke algebras and harmonic analysis on finite groups. Rend. Mat. Appl. **33**(1–2), 27–51 (2013)
61. F. Scarabotti, F. Tolli, Induced representations and harmonic analysis on finite groups. Monatsh. Math. **181**(4), 937–965 (2016)
62. L.L. Scott, Some properties of character products. J. Algebra **45**(2), 259–265 (1977)
63. L.L. Scott, Collaborations. http://people.virginia.edu/~lls2l/collaborators.htm
64. A. Sergeev, Projective Schur functions as bispherical functions on certain homogeneous superspaces, in *The Orbit Method in Geometry and Physics* (Marseille, 2000). Progress in Mathematics, vol. 213 (Birkhäuser, Boston, 2003), pp. 421–443

65. B. Simon, *Representations of Finite and Compact Groups* (American Mathematical Society, Providence, 1996)
66. S. Sternberg, *Group Theory and Physics* (Cambridge University Press, Cambridge, 1994)
67. D. Stanton, An introduction to group representations and orthogonal polynomials, in *Orthogonal Polynomials*, ed. by P. Nevai (Kluwer Academic, Dordrecht, 1990), pp. 419–433
68. J.R. Stembridge, On Schur's Q-functions and the primitive idempotents of a commutative Hecke algebra. J. Algebraic Combin. **1**(1), 71–95 (1992)
69. A. Terras, in *Fourier Analysis on Finite Groups and Applications*. London Mathematical Society Student Texts, vol. 43 (Cambridge University Press, Cambridge, 1999)
70. E.B. Vinberg, Commutative homogeneous spaces and co-isotropic symplectic actions. Russ. Math. Surv. **56**(1), 1–60 (2001)
71. J.A.Wolf, in *Harmonic Analysis on Commutative Spaces*. Mathematical Surveys and Monographs, vol. 142 (American Mathematical Society, Providence, 2007)
72. O. Yakimova, Principal Gelfand pairs. Transform. Groups **11**(2), 305–335 (2006)
73. T. Yokonuma, Sur le commutant d'une représentation d'un groupe de Chevalley fini. J. Fac. Sci. Univ. Tokyo Sect. I **15**, 115–129 (1968)
74. T. Yokonuma, Sur le commutant d'une représentation d'un groupe de Chevalley fini. II. J. Fac. Sci. Univ. Tokyo Sect. I **16**, 65–81 (1969)

Index

T. Ceccherini-Silberstein et al., *Gelfand Triples and Their Hecke Algebras*, Lecture Notes in Mathematics 2267, https://doi.org/10.1007/978-3-030-51607-9

LECTURE NOTES IN MATHEMATICS Springer

Editorial Policy

1. Lecture Notes aim to report new developments in all areas of mathematics and their applications – quickly, informally and at a high level. Mathematical texts analysing new developments in modelling and numerical simulation are welcome.

   Manuscripts should be reasonably self-contained and rounded off. Thus they may, and often will, present not only results of the author but also related work by other people. They may be based on specialised lecture courses. Furthermore, the manuscripts should provide sufficient motivation, examples and applications. This clearly distinguishes Lecture Notes from journal articles or technical reports which normally are very concise. Articles intended for a journal but too long to be accepted by most journals, usually do not have this "lecture notes" character. For similar reasons it is unusual for doctoral theses to be accepted for the Lecture Notes series, though habilitation theses may be appropriate.

2. Besides monographs, multi-author manuscripts resulting from SUMMER SCHOOLS or similar INTENSIVE COURSES are welcome, provided their objective was held to present an active mathematical topic to an audience at the beginning or intermediate graduate level (a list of participants should be provided).

   The resulting manuscript should not be just a collection of course notes, but should require advance planning and coordination among the main lecturers. The subject matter should dictate the structure of the book. This structure should be motivated and explained in a scientific introduction, and the notation, references, index and formulation of results should be, if possible, unified by the editors. Each contribution should have an abstract and an introduction referring to the other contributions. In other words, more preparatory work must go into a multi-authored volume than simply assembling a disparate collection of papers, communicated at the event.

3. Manuscripts should be submitted either online at www.editorialmanager.com/lnm to Springer's mathematics editorial in Heidelberg, or electronically to one of the series editors. Authors should be aware that incomplete or insufficiently close-to-final manuscripts almost always result in longer refereeing times and nevertheless unclear referees' recommendations, making further refereeing of a final draft necessary. The strict minimum amount of material that will be considered should include a detailed outline describing the planned contents of each chapter, a bibliography and several sample chapters. Parallel submission of a manuscript to another publisher while under consideration for LNM is not acceptable and can lead to rejection.

4. In general, **monographs** will be sent out to at least 2 external referees for evaluation.

   A final decision to publish can be made only on the basis of the complete manuscript, however a refereeing process leading to a preliminary decision can be based on a pre-final or incomplete manuscript.

   Volume Editors of **multi-author works** are expected to arrange for the refereeing, to the usual scientific standards, of the individual contributions. If the resulting reports can be forwarded to the LNM Editorial Board, this is very helpful. If no reports are forwarded or if other questions remain unclear in respect of homogeneity etc, the series editors may wish to consult external referees for an overall evaluation of the volume.

5. Manuscripts should in general be submitted in English. Final manuscripts should contain at least 100 pages of mathematical text and should always include

   - a table of contents;
   - an informative introduction, with adequate motivation and perhaps some historical remarks: it should be accessible to a reader not intimately familiar with the topic treated;
   - a subject index: as a rule this is genuinely helpful for the reader.
   - For evaluation purposes, manuscripts should be submitted as pdf files.

6. Careful preparation of the manuscripts will help keep production time short besides ensuring satisfactory appearance of the finished book in print and online. After acceptance of the manuscript authors will be asked to prepare the final LaTeX source files (see LaTeX templates online: https://www.springer.com/gb/authors-editors/book-authors-editors/manuscriptpreparation/5636) plus the corresponding pdf- or zipped ps-file. The LaTeX source files are essential for producing the full-text online version of the book, see http://link.springer.com/bookseries/304 for the existing online volumes of LNM). The technical production of a Lecture Notes volume takes approximately 12 weeks. Additional instructions, if necessary, are available on request from lnm@springer.com.

7. Authors receive a total of 30 free copies of their volume and free access to their book on SpringerLink, but no royalties. They are entitled to a discount of 33.3 % on the price of Springer books purchased for their personal use, if ordering directly from Springer.

8. Commitment to publish is made by a *Publishing Agreement*; contributing authors of multiauthor books are requested to sign a *Consent to Publish form*. Springer-Verlag registers the copyright for each volume. Authors are free to reuse material contained in their LNM volumes in later publications: a brief written (or e-mail) request for formal permission is sufficient.

Addresses:

Professor Jean-Michel Morel, CMLA, École Normale Supérieure de Cachan, France
E-mail: moreljeanmichel@gmail.com

Professor Bernard Teissier, Equipe Géométrie et Dynamique,
Institut de Mathématiques de Jussieu – Paris Rive Gauche, Paris, France
E-mail: bernard.teissier@imj-prg.fr

Springer: Ute McCrory, Mathematics, Heidelberg, Germany,
E-mail: lnm@springer.com

Printed in the United States
By Bookmasters